S0-BNK-465

THE LIVING STATE
With Observations on Cancer

THE LIVING STATE

With Observations on Cancer) 1972.

ALBERT SZENT-GYÖRGYI

Laboratory of the Institute for Muscle Research
Marine Biological Laboratory
Woods Hole, Massachusetts

 1972

ACADEMIC PRESS   New York and London

*To the memory of
my wife Márta and my daughter Nelly*

CONTENTS

PREFACE

It would take a superhuman ability to explore the deeper foundations of life. What one can do, more hopefully, is to investigate just a few of its facets. This booklet is a very personal account of some of my own researches. It is neither complete nor continuous. I have omitted even much of my own work. No mention is made, for instance, of my work on metabolism, the C_4 dicarboxylic acid catalysis which was honored by the Nobel Prize and which has led to the Krebs cycle.

References will be found at the end of each chapter. I have tried to summarize the work of authors who have done extensive research.

In its early phases my research was generously supported by the National Institutes of Health. My thanks are due to all those who have helped to keep me "above water" since, particularly the Josephine B. Crane Foundation. My thanks are also due to the National Science Foundation (Grant GB 29395) and to the L. and L. Foundation. I gratefully acknowledge the assistance of my faithful associates Miss Jane A. McLaughlin and Miss Barbara Perry, and of Dr. L. S. Együd. I also want to thank my friend Dr. John Platt for his painstaking criticism. Last, but not least, I owe a debt of gratitude to the Marine Biological Laboratory, Woods Hole for providing me with a scientific home.

Albert Szent-Györgyi

I

INTRODUCTION

Every biologist has at some time asked "What is life?" and none has ever given a satisfactory answer. Science is built on the premise that Nature answers intelligent questions intelligently; so if no answer exists, there must be something wrong with the question. Life, as such, does not exist. What we can see and measure are material systems which have the wonderful quality of "being alive." What we can ask more hopefully is "What are the properties which bring matter to life?"

Though I do not know what life is, I have no doubt as to whether my dog is alive or dead. We know life by the existence of things for which there is no direct physical reason and which even seem contrary to the rules of physics. Life appears to be a revolt against the rules of

1

Nature. It resembles the anarchistic conspiracy of Chesterton (1930) which was aimed at the abolition of all rules, but had rules of its own, stricter than the ones against which it revolted. Life is a paradox. It is easy to understand why man always divided his world into "animate" and "inanimate," *anima* meaning a soul, the presence of which had to explain queer behavior.

The most basic rule of inanimate nature is that it tends toward equilbrium which is at the maximum of entropy and the minimum of free energy. As shown so delightfully by Schrödinger in his little book, "What is Life" (1945), the main characteristic of life is that it tends to decrease its entropy. It also tends to increase its free energy. Maximum entropy means complete randomness, disorder. Life is made possible by order, structure, a pattern, which is the opposite of entropy. This pattern is our chief possession, it was developed over billions of years. The main aim of our individual existence is its conservation and transmission.

Pattern and structure can turn things around. As a rule, opposite charges approach, neutralize one another, and produce motion. Within structures, such as that of a dynamo, it is motion which separates charges. Life took its own course when its first pattern was established. Life is a revolt against the statistical rules of physics. Death means that the revolt subsided and statistical laws resumed their sway.

Trying to approach life we must bear these relations in mind to avoid acting as Tyndall's chemist did, who, when asked to find out what a dynamo was, dissolved it in hydrochloric acid. So when trying to understand life we must bear pattern in mind, the specific relations being summed up by the word "organization" which means that the whole is more than the sum of its parts, $2 + 2 > 4$, which is the basic mathematical equation of biology. When dissolving the system into its parts, we end up with 2's having lost the $>$.

In arranging atoms into molecules, say, protein molecules, Nature passes through three stages of organization. First she produces fibers, a linear array, arranging the atoms relative to one another to achieve the proper "configuration." Then she folds up the fiber to achieve a specific "conformation." Eventually she brings the various molecules into a specific relation relative to one another, establishing "coordination." The molecules, thus ordered, may influence and alter each other's properties and produce reactions even over distances.

When the chemist or physicist wants to study the interaction of two particles, say, two molecules, he tries to isolate them to avoid outside interference. Should he want to transfer, for example, an electron from molecule A to molecule B

$$A + B \rightleftharpoons A^+ + B^-$$

he will probably calculate the energy needed to separate a plus and minus charge, calculate ionization potentials and electron affinities. But life is mostly the result of a series, a chain of reactions, and knowledge of a single reaction has but limited value. So if A and B are members of an electron-transfer chain, A, while transfering an electron to B, may receive a new electron which will abolish its plus charge. So there will be no force to pull the electron back from B, and electrons may smoothly flow through the chain, from A to B, from B to C, etc.

An important difference between pure physics and biophysics is in probability.* While physics is the sci-

*Probability has a deep physical meaning. The most probable state of the universe is that of a minimum of free energy and maximum of entropy, randomness. This is the state toward which

ence of the probable, biology is, in a way, the science of the improbable. Probable chemical reactions occur spontaneously. If biological reactions were "probable," they would take place spontaneously, and we would burn up; our machine would run down as a watch relieved of its regulation. Life, on principle, has to work with improbable reactions which it then makes proceed by specific routes, thereby regulating them. Life, altogether, is an improbable phenomenon which was generated, perhaps, but once during the billions of years of the history of

---

the universe tends, which makes time flow in its present direction. Once this state is reached, there will be no life. What, then, keeps the universe from reaching this point? This has been discussed by Dyson (1971). Why do not all energy-producing reactions take place, letting the entire biosphere run down? The situation is analogous to that of a rock on the mountain side. A rock rolling down liberates great amounts of energy, and if the mountain does not crumble it is because loosening rocks demand a small amount of energy that has to be invested before the rock can start rolling; this energy is not available. The energy liberated by the rolling rock may be very great compared to the energy which has to be invested, and if there were a way to have the energy, liberated by rolling rocks, used for loosening new rocks, mountains would disappear. What keeps mountains standing is the small amount of initial investment.

Similarly, to make a chemical reaction proceed, as a rule, a relatively small amount of "activation energy" has to be invested. This is what keeps the biosphere from running down. What life does is to decrease the necessary initial investment by means of enzymes, till the small amount of energy supplied by heat agitation becomes sufficient to make the reaction proceed. Life, then, conserves the energy liberated to initiate new reactions. From an energy point of view life rests on three pillars. One is the decrease of activation energy, accomplished by enzymes. The second is the conservation of the energy liberated and its investment in new reactions. This is done by means of the "high energy phosphates," $\sim P$'s. The third is photosynthesis, by means of which life uses the radiation energy of the sun to build new molecules from which energy can be liberated.

the world. If I were to ask a physicist what the probability was that the trillions of electrons and atomic nuclei would get together and stay in the relative position they are in me, the answer would be that the probability was practically zero, which means that I am impossible. One of the main aims of biology is to find out the way in which life makes reactions proceed, thereby perfecting itself.

When pursuing this analysis we must be careful not to kid ourselves into believing that we understand, when we do not. Analyzing living systems we often have to pull them to pieces, decompose complex biological happenings into single reactions. The smaller and simpler the system we study, the more it will satisfy the rules of physics and chemistry, the more we will "understand" it, but also the less "alive" it will be. So when we have broken down living systems to molecules and analyzed their behavior we may kid outselves into believing that we know what life is, forgetting that molecules have no life at all.

We must also be very careful of how we ask questions, for by the way we ask them we may determine the answer. If we ask Nature "Is light a particle?" Nature will answer "Yes, it is a particle." But if we ask "Is it a wave?" Nature will answer "It is a wave." Is a gramaphone playing a Bach record a purely physical system? The answer is "Yes it is." The needle follows the groove and the membrane follows the needle. The only thing I left out was the genius of Bach without which the whole thing would make no sense. My watch too, is a purely physical system, but I should not forget the generations of watchmakers who have developed this wonderful little gadget which could never have come together by random fluctuations.

Every action must have its underlying mechanism, and a system can only do what its structure allows it to do. A record player will never write a letter, a typewriter will never make music, and a cow will never be able to lay an egg, however hard she tries. But once the system is there it will be able to do what its structure allows it to do. So structure and function are, in a way, identical, and we may study either or, more correctly, must study both on all levels. They are one. Structure generates function, function generates structure. For thousands of years man has observed living structures on the macroscopic level. For a century man has studied them on the microscopic level. At present we are concentrating on the molecular level, below which the electronic level awaits exploration, the road having been paved by the Pullmans.

Biochemistry and biophysics are very young sciences, still in their baby shoes. It was less than seven decades ago (1903) that Büchner discovered that cell-free filtrates of yeast could ferment sugar, which meant that biological reactions could be separated from life and be analyzed. When my beloved teacher Sir Frederick Gowland Hopkins had set out to study the production of lactic acid in muscle contraction he was told that biochemistry is a "*contradictio ad absurdum*," an impossibility, because as soon as we break down the system it is not alive any more.

We have penetrated deeply into cellular mechanisms and begin even to ask questions about the origin of life. In my opinion the study of the cell can lead us a long way in that direction too. Life has developed its processes gradually, never rejecting what it has built, but building over what has already taken place. As a result, the cell resembles the site of an archeological excavation with the successive strata on top of one another, the oldest one the deepest. The older a process, the more basic a role it plays and the stronger it will be anchored, the newest processes being dispensed with most easily. This is why in division the cell reverts to simpler, more archaic processes of energy production.

Where biochemistry and biophysics will lead us nobody can predict. They may change our life more than all the other sciences already have.

My own scientific career was a descent from higher to lower dimension, led by the desire to understand life. I went from animals to cells, from cells to bacteria, from bacteria to molecules, from molecules to electrons. The story had its irony, for molecules and electrons have no life at all. On my way life ran out between my fingers. The present book is the result of my effort to find my way back again, climbing up the same ladder I so laboriously descended. Having started in medicine, it is befitting that I should end with a medical problem, cancer, which took away most of what was dear to me.

REFERENCES

Chesterton, G. K. (1930). "Man Who Was Thursday." Arrowsmith.

Dyson, F. J. (1971). *Sci. Amer.* **224**, 50.

Pullman, B., and Pullman, A. (1963). "Quantum Biochemistry." Wiley (Interscience), New York.

Schrödinger, E. (1945). "What is Life?" Macmillan, New York.

II

WATER, THE HUB OF LIFE

Life originated in the ocean. Water is its *mater* and *matrix*, mother and medium. Life could leave the ocean when it learned to grow a skin and take the water with it. We are still water animals, only we have the water inside, not outside. We are a walking aquarium. We are a 20% aqueous solution. We have guarded our legacy, the water of the sea, carefully. The ionic concentrations of our blood still reflect the ionic concentrations of the primordial ocean. It is a remarkable fact that we have probably kept these concentrations more constant than the ocean itself, which has since changed.

When a biochemist studies a system his first question is "What is it made of?" Life, having originated in the ocean, could build its machinery only from what it found there. It found three things: water, organic molecules (including CO_2), and ions. No structures could be built of ions. With their charge and motility they could be used only as triggers, or for balancing electric or osmotic differences. Water, as we know it from everyday experience, is a shapeless, unreactive, neutral liquid. So, according to biology today, life's machinery is built of organic matter using water as its medium.

Sixty years of research has taught me to look upon water as part and parcel of the living machinery, if not the hub of life. Water is the most extraordinary substance! Practically all its properties are anomalous, which enabled life to use it as building material for its machinery. Life is water dancing to the tune of solids. That biologists forgot about it is no more than natural. According to Sir Oliver Lodge the last thing a deep sea fish could discover is water.

The extraordinary nature of water is borne out by the two constants used most frequently for the characterization of substances: melting and boiling points. According to the size of its molecules, water should boil at $0°C$. It boils at $100°C$. It should melt at $-100°C$. It melts at $0°C$, indicating that the water molecules tend to stick together. We have to decrease the temperature only by $1/273$, cooling it from $273°K$ to $272°K$, and water turns into a solid which can split rocks. Eskimos build their houses with it.

The unique shape of the water molecule is roughly represented in Fig. 1 (Horne, 1969). Being essentially a cloud of electrons, its outlines are not as sharp as suggested by the figure. In rapid rotation it would appear as a ball of about 3.8 Å. This queer, four-legged creature has a bend in the middle. Two of its legs are formed by its two H's, which give to their half of the molecule a positive charge enabling it to act as H donor forming two hydrogen bonds with other water molecules. The other two legs are formed by the nonbonded electron pairs of oxygen which lend to this end a negative charge, making the molecule a strong dipole, with a tetrahedral shape, enabling it to form two H bonds as acceptor with two other water molecules or whatever H donors there may be. The water molecule is thus capable of forming four H bonds of considerable strength of about 6 calories each. Each water molecule can thus bind four water molecules, each of which, in turn, can also bind four molecules. This entails that below $0°C$ water as a whole turns into one single continuous solid. The dipole character of the molecules may lead to more extensive structures. In many ways, the situation is analogous to a weak magnet which, if placed in iron filings, will polarize the nearby particles which, in turn, will polarize their neighbors, thus enabling extensive structures to be built up along the magnetic lines. In the case of water these structures may be stabilized by the H bond formation. The building up and breaking down of these structures will have a cooperative character.

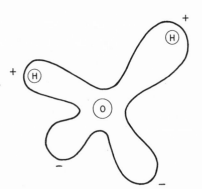

Fig. 1. Shape of the water molecule. Redrawn from R. A. Horne's "Marine Chemistry" (Horne, 1969).

Removal of one particle may lead to the collapse of the whole structure.

The cooperative behavior of water is clearly shown by the melting of ice. Though the four H bonds of the single molecules, in a mole of ice, have about 24 calories of bond energy, 1.5 calories of heat will suffice to make the ice melt, make its structure disintegrate. The water of 0°C, thus formed, still has a low specific weight, indicating that the ice did not dissolve altogether, it only broke down into smaller pieces. It was thirty years ago that Bernal and Fowler (1964) gave strong evidence that even at room temperature water is to a great extent crystalline, the single water molecules linking to long rows forming all sorts of figures. Heat agitation destroys these very fast, in 10^{-11} seconds or so, which has led to the "flickering water model" of Frank and Wen (1957), which now has wide acceptance. Water can thus form and destroy extensive structures simultaneously with great ease and speed.

These relations render most properties of water anomalous. Whatever property of water is measured we usually find a breaking point around 15°C, another around 45°C, and one in between, around 30°C (Drost-Hansen, 1956–1966). It seems likely that it is these breaking points which set the limits of life for warm-blooded organisms. If our body cools below 15°C, death ensues. Our normal body temperature may be 37°C because Nature not only wants to enjoy the advantages of heat agitation but also wants to keep at a safe distance from 45°C. If our temperature, in an infection, rises to 42°C, it is time to call the undertaker.

Water has a poorly balanced electron structure, is very electropolar, so it attracts other electropolar, poorly balanced substances or atomic groups, making them water soluble. Homopolar substances of well-balanced electronic structure are insoluble in water.

What is important to the biologist is not so much the structures formed in the bulk of water but the structures formed around solids. That the structure of solids near surfaces is different from the bulk has been suspected for a long time. Sir William Hardy (1931; Hardy and Bircumshaw, 1925) spoke about a "fourth state of matter" by which he meant the state of matter near solids. He was chiefly interested in the structures formed by lubricants near moving metal surfaces. Lubrication is due to these structures that protect the metal, so that metal will not rub against metal but lubricant against lubricant. In 1933, Derjaguin rotated a convex glass close to a flat one and found that water, between the two, behaved as a solid and showed structure elasticity if the distance between them was less than 900 Å. Over 1500 Å the water behaved as the bulk. Demény and myself (1972) measured the force needed to pull two well-polished glass plates alongside one another at a low and constant velocity. Our preliminary results are summed up in Fig. 2. At a distance of 2000 Å the force was considerable, 600 gm/cm^2. Each glass plate was thus covered by a 1000 Å deep sheet of water which behaved almost as a solid. Glass does not form H bonds so this action on the solvent could not be specific for water. p-Xylene gave similar results, and 35% sulfuric acid (circles, Fig. 2) made no difference. Water can thus not only separate two surfaces but can also hold them strongly together. This is, in a way, common knowledge. This is why we lick our fingers when turning a page in a book, or spit into our palms when we want to hold an ax more firmly. Everybody knows that one can walk on wet sand without sinking into it. When the two glass plates in our experiments were pulled alongside one another slowly (0.3 mm per minute), the force needed for the pull increased and decreased periodically. When

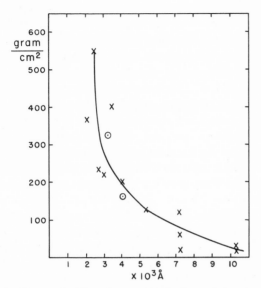

Fig. 2. Forces holding two wetted glass plates together at varied distances (L. Demény and A. Szent-Györgyi).

one pulls one's wetted finger on the rim of a glass, this periodic change produces a musical sound. Mozart tried to utilize this for building a new musical instrument.

The situation can be expected to differ in the case of surfaces capable of forming H bonds with water, as is the case with proteins. In this case the results of the interaction have to depend to a great extent on geometry. If the distribution of the H bonds, formed between water and protein, corresponds to the lattice constants of water, then long chains of water molecules may be built up which, supporting one another, build extensive structures which reach deep into the bulk of the fluid. Contrary to this, if the two geometries do not correspond, no such structures will be formed, and the surface will be covered only by one or two sheets of water molecules. While one surface may thus build up deep water structures, another may build none at all, and a structure formed may suddenly collapse if a change in the protein disfavors H bonding.

As proteins may stabilize water structures, so water may stabilize protein structures, depolarize them, separate them, or bind them together. If the water structures formed on two surfaces are confluent they will hold these surfaces together, forming a *"water bond"* between them. The water structures thus formed can be expected to have specific properties, and may have a decisive influence on the folding of fibrous protein molecules which determines their structure and function. The structured water sheets around the protein may behave as a special phase with special properties and functions, which may complement the function of the protein itself. Icelike structure induces a high proton conductivity. According to Klotz (1970), structured water may even conduct electrons or OH ions. At higher temperatures all these water structures will melt away, inducing protein denaturation.

The unique physical state of living tissues may be due, to a great extent, to water bonds. My skin is very tough; with a little tanning it would make good boots. At the same time, it is also soft and pliable. This indicates that the system is held together by a great number of weak bonds, easily formed and broken. Fewer strong bonds would make the system brittle. The water bond is a weak one, but many weak bonds are just as strong as a few strong ones. A hundred bonds of 1 calorie are as strong as one of 100. This, however, is true only if we pull perpendicular to the surface. If we start pulling at the edge, we will have to break the bonds one by one, and so will need only a force corresponding to 1 calorie, if the single bonds have this bond energy. This explains the great pliability of systems held together by weak bonds, like water bonds, the single links being broken and formed with great ease.

The ability of protein structures to bind water, hydrate, and show a rubberlike elasticity may be very intimately related to the very nature of the living state and its gradual fading, aging. The weakening of the water bonds can be expected to lead to the expansion of tissue and may be responsible for wrinkles of the skin, so typical of senescence. It may be responsible also for the widening and lengthening of veins. Unprotected by water, surfaces may also show hysteresis, stick together more easily, which may be responsible for the stiffness accompanying old age. It is possible to make a sophisticated guess about the life expectancy of an individual by pinching his skin on the back of his hand, producing a little ridge. The speed with which this ridge disappears on release is proportional to life expectancy, indicating the degree of hydration which seems to be a decisive factor of longevity and opens a fascinating field for research.

The water structures formed around surfaces may thus play a major role in the physical state of animal tissues. They may also play an important role in function. As I will show later they play a central role in muscle contraction. The failure of water to protect structures may underlie many important pathological reactions and may be responsible for the majority of human deaths. As shown by Sir William Hardy (1931), lubricants protect moving metallic surfaces by forming protective multilayers around them so that it will not be the metal which rubs against metal, but lubricant against lubricant. The situation, in articulations, may be analogous to that in lubricated metallic surfaces, the bound water forming protective sheets rubbing against one another. The chondroitic acid of cartilage with its sulfate group can be expected to be very hydrophylous. Also strongly hydrated solutes, as glucoproteins or hyaluronic acid, may act as lubricants. If, for any reason, the protein surface becomes damaged, the water structure will be damaged too and will protect no more, leading to erosion, pain, and immobility, as is the case in arthritis.

Similarly to joints, the surface of the blood vessels and heart valves needs protection against the rubbing of the circulating blood, as the blood needs protection against the rubbing of the walls of the blood vessels. Damage to the protein surface will lead to collapse of the protecting water sheet and deposition of fibrin, with ensuing stroke, heart attack, or failure of heart valves.

We can thus sum up by saying that water structures are just as much a part of the living machinery as its organic and more solid parts. The solid induces and stabilizes the water structures; the water structure depolarizes, protects, separates, or links solid structures. It is a matter of semantics whether we say that solids bind water or that water binds solids. It is hoped that the low-temperature electron microscopy of H. Fernández-Morán will soon make water structures visible.

Water not only plays an important role as part of the solid machinery, but also plays a central role in energetics. The driving force of life is the energy of solar radiation which is conserved by being used to separate the elements of water, H and O, or by taking a water molecule from between two phosphate molecules (Arnon's cyclic phosphorylation). The energy thus stored can later be utilized by reversing these processes and allowing the H and O to unite again (biological oxidation) or by putting the water molecule back between the phosphates (hydrolysis of $\sim$P, the high energy phosphate bonds). In both processes water plays a central role. It is the hub of bioenergetics. As Rybak puts it (1968), "bioenergetics is but a special aspect of water chemistry."

Whatever we study in biochemistry we find that water plays an essential role, not only as such (H_2O), but also in its ionic state (H^+ and OH^-), determining pH, one of the most basic parameters of life. Life processes also involve, to a great extent, a shift of electrons (e.g., electron flow in oxidation or photosynthesis). Electron flow must be accompanied by a flow of protons, H^+, to maintain electroneutrality. The protons possibly move in the structured water surrounding the protein particles. Even apolar, hydrophobous molecules build crystalline water structures (clathrates) around themselves.

Substances are chiefly built-up or broken down by taking out or putting in H_2O molecules (hydrolysis and synthesis). Biological oxidation (as will be discussed later) is, as a rule, not a coupling with O_2, but simply a replacement of H's by the water, H and OH, which makes the substrate gradually richer in O till eventually only CO_2 and H_2O remain. Oxygen comes in only as a final electron acceptor.

All this may be common knowledge. I mention it because we tend to concentrate only on the substances to be split, joined, or oxidized and forget the molecule which plays the central role in all these processes, water (Szent-Györgyi, 1971).

REFERENCES

Bernal, J. D., and Fowler, R. H. (1964). *J. Chem. Phys.* **40**, 2800.
Demény, L., and Szent-Györgyi, A. (1972). Paper in preparation.
Derjaguin, B. V. (1933). *Z. Phys.* **84**, 657.
Drost-Hansen, W. (1956-1966). *Ann. N. Y. Acad. Sci.* **125**, 471.

Drost-Hansen, W. (1965). *Ind. Eng. Chem.* 57, 38 (March), 18 (April).

Frank, H. S., and Wen, W. Y. (1957). *Discuss. Faraday Soc.* 24, 133.

Hardy, W. (1931). *Phil. Trans. Roy. Soc. London, Ser. A* 230, 1.

Hardy, W., and Bircumshaw, L. (1925). *Proc. Roy. Soc., Ser. A* 108, 2.

Horne, R. A. (1969). "Marine Chemistry." Wiley (Interscience), New York.

Klotz, I. M. (1970). *In* "Membranes and Ion Transport" (E. Bittar, ed.), Vol. 1, p. 93. Wiley (Interscience), New York.

Rybak, B. (1968). "Principles of Zoophysiology," Int. Ser. Monogr. Pergamon, Oxford.

Szent-Györgyi, A. (1971). *Perspect. Biol. Med.* 14, 239.

III

MOTION AND MUSCLE

It was more than thirty years ago that I felt I had enough research experience to attack a more difficult biological problem associated with what we call "life." We know life from death by what it does. Most signs of life are energy transductions, transduction of chemical energy into some sort of work, mechanical, electrical, or osmotic, the products being motion, reflexes, or secretions. Of these products motion is the oldest, the most obvious, and apparently the simplest. The organ of motion is muscle. So I started working on muscle, and spent twenty years trying to understand it.

We move because our muscles contract, shorten, their two ends approaching one another. Since motion has to be produced on the molecular level, we could start by

19

asking how could two points, A and B, be made to approach one another on this level? The simplest way to do this would be to connect them by means of a molecular filament (Fig. 3a), which tends to shorten. On releasing the thread it would contract, shorten, bringing A and B closer together (Fig. 3b). There are various ways in which molecular threads could fold and shorten, but whatever the mechanism or driving force of folding may be, hydration, a stiff water jacket, could keep them straight, as space suits keep our astronauts stiff. Folding, shortening, and contraction could then be induced by shedding the water jacket. In Fig. 3 a hydration is symbolized by shading. A different kind of motion could be produced by using two threads, or rods, linked together, of which only one is contractile (Fig. 3c). If two metal rods were soldered together and only one of these would shorten when cooled, the doubled rod would have to bend toward the side which contracted. Such a system is called a "bimetallic strip." Many thermometers and barometers are built on this principle, and such a dual system is often called a "bimetallic strip," regardless of whether its two parts are actually metal. Thus if two molecular filaments were linked together side by side, and only one of them shortened, contraction of this filament would lead to a bending (Fig. 3d). If we were to twist this "bimetal" this shortening would lead to the formation of a corkscrew (Fig. 3e). If the whole system were not dehydrated at one time, but dehydrated at one point only and proceed through the whole thread, then a wave motion would be generated, and a "corkscrew" would perform a screw motion.

When I embarked on muscle research my first problem was: What to do? How to start? There is one thing one can always do: repeat what others have done before

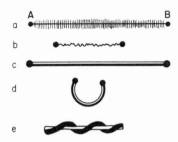

Fig. 3. Folding on the molecular level. For explanation see text.

(A. Szent-Györgyi, 1945, 1947). If one keeps one's eyes and mind open, one may observe something new that others missed and this can put one on a fresh trail. So my faithful associate Banga and I (1941-1942) started with repeating what Willy Kühne (1863) did a century earlier: we minced rabbit muscle and suspended it in strong (0.6 M) KCl solution, and, as described by Kühne, out came a big quantity of the viscous protein which he called "myosin." Edsall (1930) and Muralt and Edsall (1930) showed later that it consisted of elongated particles. Evidently, myosin was a contractile protein. The trouble was that it would do nothing *in vitro*. A contractile protein should contract wherever it is. Evidently, it was missing something.

One day we observed that if we stored our minced muscle, suspended in KCl, it became more sticky. This could be due either to more myosin or a new protein. It was a new protein. We called it "actin." It was isolated in brilliant fashion by my young associate F. B. Straub (1942, 1943).

While my laboratory was doing this work, Engelhardt and Ljubimova (1939) discovered that myosin could act as ATPase, splitting off ATP's terminal phosphate. At that time the idea that this splitting could supply the energy for biological processes began to emerge. We found myosin, in itself, to be a very poor ATPase, but its enzymic activity was greatly increased by actin. One protein could thus change the enzymic properties of another, a rather surprising and new observation! Actin changed not only the enzymic properties of myosin, it changed its whole physical state, the system becoming more solid, viscous, or resilient. Actin and myosin formed a highly viscous complex we called "actomyosin." We could also show that resting muscle contained no actomyosin, but behaved as if it contained only actin and myosin side by side. It was only during an excitation, in the "active state," that actomyosin was formed. This is why, at rest, muscle is soft and does not split its ATP.

The most striking property of actomyosin was its enormous affinity for water, its strong hydration. I never succeeded in making a stronger actomyosin solution than 3%, containing 97% water. At a physiological salt concentration actomyosin is a gel, and H. H. Weber (1934), a pioneer of muscle research, has taught us how to make threads of it by dissolving it in a strong salt solution and squirting this solution in a thin jet into water. The water elutes the salt, and the protein precipitates in the form of threads. But the threads would not move, either; something was still missing, possibly some smaller molecule. So we prepared a *Kochsaft*, a hot watery extract from muscle, and suspended our threads in it. They contracted, shortened. Seeing motion, this age-old sign of life reproduced *in vitro* for the first time, was the most exciting moment of my long scientific life. It took only a little cookery to show that what made the threads contract was ATP, which could supply the energy needed for contraction. That what we saw was not merely colloidal shrinking, syneresis, but a real contraction, shortening, could be shown by preparing threads in which the actomyosin filaments were arranged parallel to the axis (which could be done by stretching). On addition of ATP such a thread shortened *and* became wider, contracted without losing volume, as muscle does. Only when the filaments were distributed at random did they make the threads shorter *and* thinner by their contraction. We could thus say that our threads contracted because the actomyosin particles shortened.

To bring the analogy of actomyosin and muscle still closer, I searched for a muscle from which long single fibers could be prepared. I found it in the musculus psoas of the rabbit which, since, has become a classic material in muscle research. Then I immersed fiber bundles of this muscle in 50% glycerol (A. Szent-Györgyi, 1949) and stored the "glycerinated" muscle in the deep freeze (−20°C). Suspended in physiological saline, at room temperature, on addition of ATP the fibers not only contracted, they developed the same tension as they developed maximally *in vivo*. This method of glycerination was used later in England for the conservation of sperm, and plays, today, an important role not only in muscle physiology but also in animal husbandry and in artifical insemination of humans. Glycerol destroys the excitatory mechanism of muscle while leaving the contractile mechanism unharmed, so the glycerinated muscle supported my statement that *muscle contraction essentially is an interaction of actomysin with ATP.*

What impressed me most in contracting actomyosin threads was their enormous changes in hydration. In contraction the strongly hydrated threads became completely anhydrous, retaining only the water which could be expected to be present as an inclusion. I had never any doubt that this *change in hydration was the central happening of muscle contraction.* This dehydration could be domonstrated not only with actomyosin threads, but also with actomyosin suspensions which, on addition of ATP, precipitated violently. The floccules contracted, settling rapidly to the bottom of the test tube in the form of a dense little plug. This we called "superprecipitation." It is a most striking phenomenon. If the actomyosin suspension is not broken into micelles it contracts to a small plug. All this has left little doubt in my mind that *muscle contraction is essentially a play of water, the building and destruction of water structures built up by the contractile protein.* Very little actin was needed to enable myosin to produce these changes in hydration, so, evidently, it had to be the myosin which bound and released the water. Actin had only a catalytic action.

This was how far I could get with the methods available at that time. There was no electron microscope as yet and X-ray diffraction had only started to invade biology.

The first results obtained with the electron microscope, in the hands of H. E. Huxley (1963) and A. F. Huxley (1957), seemed to show that all my assumptions were wrong. Myosin, in muscle, was present in the form of "thick filaments," while actin formed "thin filaments" which surrounded the thick filaments hexagonally. In contraction none of them shortened. They simply slid alongside one another (Fig. 4). Since the single muscle fibers are cut up by the cross "Z membranes" into partitions or "sarcomeres," this sliding made the sarcomeres, and, with it, the whole muscle contract. There was no shortening of filaments. This is the "sliding model" of muscle contraction which

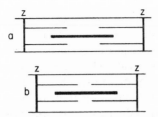

Fig. 4. Schematic representation of the sliding model of muscle contraction. a, Relaxed sarcomere; b, contracted sarcomere. The thick filaments are myosin, the thin ones actin. Z's are Z membranes.

seemed to make all my previous observations look senseless.

In the Institute for Muscle Research, Woods Hole, E. Mihályi and A. G. Szent-Györgyi (1953) pursued further an observation made by J. Gergely. What Gergely found was that the myosin molecule could with trypsin be broken down into smaller particles without losing its ATPase activity. Two kinds of subunits, "meromyosins," were produced. One sedimented in the ultracentrifuge more slowly and was called "light," "L meromyosin" (LM), while the other sedimented faster and was called "heavy," "H meromyosin" (HM). The later studies of A. G. Szent-Györgyi (1952, 1953) showed that it was only the HM which interacted with actin, so it had to be the HM that was responsible for binding water and the changes in dehydration. Only the HM had ATPase activity that could be activated by actin.

Thus attention turned from myosin to H meromyosin and the electron microscope, and Rice (1961, 1966) and H. E. Huxley (1963) showed that the fibrous molecule of meromyosin had a globule at one end. The whole particle consisted of a stalk and a head. The electron micrographs of Slayter and Lowey (1967) showed the stalk to be a helix formed by two filaments, each of which had a globule on one end (Fig. 5). Mueller (1966) showed that it was this globule only which interacted with actin. So, evidently, it was the stalk which was responsible for hydration, while it was the head which interacted with actin and split ATP, and, by doing so, induced the dehydration and shortening of the stalk. Before trying to relate these observations into a coherent whole I must deviate for an instant to consider a few corollary observations.

Bailey (1948) discovered that in addition to actin and myosin muscle contains a third fibrous protein, tropomyosin. This tropomyosin is attached to actin alongside. The thin "actin filaments" of muscle actually are actin-tropomyosin fibers. Ebashi and his associates discovered that attached to these actin-tropomyosin fibers there is a globular protein at regular intervals of 400 Å which they called "troponin" (Ebashi and Endo, 1968). This troponin has a great affinity for Ca^{2+} and has a repellant action toward the HM globule. According to Ebashi, the troponin communicates its repellant action over its 400 Å long stretch to the actin-tropomyosin fiber. What excitation does is to liberate Ca^{2+} from the "sarcoplasmic reticulum." This Ca^{2+}, then, is bound by the troponin, neutralizes its negative charge, and abolishes herewith its repellant action toward the HM head.

The curious phenomenon "metachromasia" explains this repulsion. "Metachromasia" means that the substance in question colors itself in certain basic dyes with a color that is different from the dye's own—colors itself, say, red, while the dye itself is blue. Changes in the color of the dye can be brought about by the association, or stacking of its molecules, and according to the theory of Bradley and Wolf (1959), metachromasia is due to such stacking. Kaminer and I (1963)

Fig. 5. Rough sketch of H meromyosin particle.

have shown tropomyosin to be metachromatic. Tropo- nin is even more strongly so. Since the basic dyes are bound by the COOH groups of the protein, these COOH groups must be stacked too, are not distributed at random but are present in groups, the single groups corresponding to a polyvalent charge. According to the rules of colloid chemistry polyvalent ions have a strong- er electrostatic action than corresponds to the number of their charges. Their outward action is proportional to the square of the number of charges, a bivalent ion acting four times as strongly as a monovalent one. The metachromasia of troponin and tropomyosin thus ex- plains their strong repellant action toward the negatively charged HM globule.

How the single parts are put together in muscle has been cleared up to a great extent recently by H. E. Huxley (1969) who found that, after all, actin and myosin were not really separated; there were "cross bridges" between the two, distributed with great regularity. These cross bridges are the H meromyosins which are attached loosely to the thick filaments that consist of L meromyosin. The H meromyosins are attached to the thick filaments by a "hinge," and have a certain mobility. They can be repelled by actin during relaxation and be attracted to it during contraction by means of their globules. Thus there *is* actomyosin in contracting muscle. This allows one to put the whole story, tentatively, together, and to correlate molecular structure with the function of muscle.

Muscle has three functional states: the relaxed, the activated, and the contracted.

Relaxed state. In this state the muscle is soft and pliable and corresponds to the top drawing (a) in Fig. 6. The thick myosin filament has an H meromyosin attached to it on either side. (Only part of the LM is represented.) The HM is straight and hydrated (symbolized by shading). There is no actomyosin, no interaction of actin and myosin. The HM is repelled by the thin filament and its troponin.

Activation. This state corresponds to the middle figure (b) in Fig. 6. The Ca^{2+} released by the sarcoplasmic reticulum was bound by the troponin, neutralized its repulsion toward the H globules which became attached to the thin actin-tropomyosin-troponin thread. Actomyosin was formed. The stalk of the H meromyosin is still straight and hydrated but its interaction with actin induced in it a new physical state, a kind of stiffness, responsible for the resilience of muscle in this state.

Contraction (Fig. 6 bottom, c). The globule of the H meromyosin, having been enzymically activated by its interaction with actin, has split the ATP attached to it, while the stalk dehydrated and contracted. This makes the thin and thick filaments slide relative to one anoth-

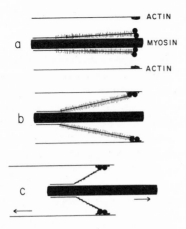

Fig. 6. Schematic representation of muscle contraction on the molecular level. a, Relaxed state; b, active state; c, contracted state.

er, causing the muscle to shorten, contract.

The relation of the splitting of ATP and contraction is an old and unsolved problem. It is possible that the actual contraction follows the splitting off of the terminal phosphate of ATP, transforming the energy of the $\sim$P into mechanical work. However, it is equally possible that the ATP, linked to the globule, produces contraction as such, the system having been activated by actin. In this case the energy, spent in contraction, is the potential energy of the system. Eventually, thermodynamic balance is established at the expense of the subsequent splitting of ATP.*,†

*In contraction the ATP (adenosine triphosphate) is dephosphorylated to ADP (adenosine diphosphate). The ADP is rephosphorylated to ATP by a very powerful enzyme, phosphoferase, which transfers the phosphate of creatine phosphate onto it. There is also a very active enzyme present which transforms two ADP's to one ATP and one AMP (adenosine monophosphate).

†After I completed this book an article appeared by Bowen and Mandelkern (1971) with new and serious arguments for the second, postenergitazation theory.

29

Though various details still have to be cleared up, this whole story is like a movie with a happy ending: everybody was correct, there *is* actomyosin, there *is* hydration and dehydration, and there *is* shortening and there *is* sliding. The role of ATP in this cycle is not limited to its splitting. Attached to the HM globule, ATP has to contribute, with its three negative charges, to the repulsion between troponin and the globule. After contraction, the rephosphorylation of the ADP to ATP contributes to the renewed repulsion between troponin and the HM globule, and thus promotes relaxation. Having released actin the stalk of HM hydrates again. The whole system having thus returned to its initial state [top (a) of Fig. 6] is ready for a new contraction. The role of ATP is not limited to the contraction cycle. It dominates the physical state of muscle even in rest, keeping it soft and pliable, keeping actomyosin dissociated. I have shown, with Borbiro (Borbiro and Szent-Györgyi, 1949), that *rigor mortis* is but a lack of ATP, the ATP having become dephosphorylated postmortem, allowing actomyosin to be formed without contraction. Various points of this cycle may still demand experimental confirmation, but the whole story fits so well together that I have no doubt of its basic correctness.

This story, if correct, brings to light a new phenomenon, which to my mind is the most amazing of biological puzzles, closely related to the very nature of life. According to the above story the whole long H meromyosin particle acquires new physical properties when the heads touch actin, and the long stalk of the H meromyosin dehydrates when its head splits ATP. How can a chemical reaction, taking place on the globule, produce such a profound change in the long stalk? That one macromolecule should be able to change the properties of another molecule over a relatively long distance is a new and striking phenomenon. This demands a specific relationship which, in the Introduction, I called "coordination." Such an action at a distance may be one of the most basic and important factors in biology.

The main actor in our story is the H meromyosin, the stalk of which contracts using the energy released by the head on splitting one ATP molecule. The contraction of the single HM's is integrated to muscular concentration by the long thin and thick filaments. All this makes muscle a rather complicated machinery. One may ask how Nature could develop such a very complex system to produce motion, one of the oldest phenomena of life.

Nature develops its structures stepwise, adding to them and improving them gradually, never throwing away the foundation. If the core of the complex structure is the H meromyosin, then it should also be its oldest and most primitive part from which Nature developed muscle. So this core, the H meromyosin, should also be found in more primitive motile organisms.

The simplest motile organisms are microorganisms, such as the vibrios or some of the bacteria. What makes them move are thin molecular threads. The vibrios have one single filament at their tail end, while the *Proteus* bacillus has many filaments around its body beating in concert (Fig. 7). The filaments produce a wavelike motion, similar to a wave traveling along a whip. By doing so they propel the body. Astbury and his associates Beighton and Weibull (1955) studied these filaments with X-rays and found them to be "monomolecular muscles." Consisting of one molecular thread only, they have no circulation, and so the energy for their motion has to be supplied at their base, which is within the bacterial body. It is difficult not to discover in these threads the analog of the stalk of the H meromyosin. I suspect that by studying these structures more carefully one will also discover a head, imbedded in the bacterial body, which liberates the energy needed for motion by splitting ATP. In muscle the stalk of meromyosin has to contract over its whole length simultaneously, which can be done by a single filament. The bacterial hairs have to produce a wave motion which demands a "bimetallic strip," the contraction progressing as a wave. Astbury and his associates actually found indication of two different proteins, one of which, the contractile, may be analogous to the stalk of HM, while the other may be analogous to any of the noncontractile proteins, actin, tropomyosin, or L meromyosin. The nature of the wave progressing through these filaments is unknown, and is one of the most intriguing of biological puzzles. What I want to emphasize here is that in these bacterial organs one can see with one's own eyes these waves, which leave no doubt about their existence.

When organisms became too big to be moved by single filaments, these filaments were collected into more complex filaments with coordinated movements, as is the case in the tails of flagellates or of sperm, or ciliae of cilliated epithelium. To move still bigger organisms, the contractile filaments had to be collected in specific organs, muscle, their motion being integrated by the long filaments.

Such waves need not necessarily be limited to filamentous organs. They might take place, for instance, in the wall of pores of membranes. The water structures, thus generated, may drive before them dissolved molecules or ions and be responsible for "active transport." This would be analogous to "zone melting" (Pfann

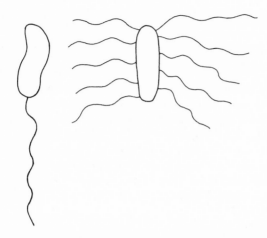

Fig. 7. Schematic representation of *Desulfovibrio desulfuricans* (left) and *Proteus vulgaris* (right).

1962, 1967).* Figure 8 is meant to be a magnified picture of a pore in a membrane, the shaded part the water structure built in the pore by the progressing wave. If pores, in membranes, are permanently filled with icelike water, then the waves could cause a collapse, a "melting" of this structure. In this case the shaded strip in Fig. 8 would stand for "molten" water, containing the transported particles. Such waves may be responsible also for "protoplasmic streaming," the most archaic analog of blood circulation. That dissolved substances can be pushed along by water structures can easily be demonstrated by freezing an aqueous dye so-

---

*After the completion of this book a paper appeared by Dragomir (1971) pointing out the possible analogy between zone melting and active transport.

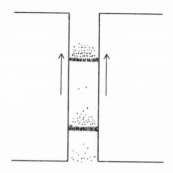

Fig. 8. Schematic representation of a pore in a membrane. Arrows indicate direction of wave; shaded lines represent water structure or melted zones.

lution (toluidine blue) in a test tube by slowly immersing it in a freezing mixture. If the solution is stirred, the whole dye will be found, eventually, in the thin sheet of water on top of the colorless ice filling the tube.*

In hydration and dehydration, as well as in the generation of waves, K and Na ions may also be involved. The contraction of actomyosin is limited to a narrow range of the concentration of these ions and, according to Ling (1962), the K ions in tissues are not free but are bound to protein. So a change in their concentration or binding may be involved in building water structures.

During the last decade the study of water structure has found a powerful new tool in nuclear magnetic resonance. Bratton, Hopkins, and J. Weinberg (1965)

---

*It seems likely to me that the first motion, in the course of evolution, was the motion of the solvent, water, driven by waves propagated within the living structures, in the surrounding water. This motion of water had to increase the chances of capturing a foodstuff molecule. It may have been much later that the structures responsible for generating the waves formed filaments, protruding into the solvent, eventually propelling the whole body.

applied NMR to contracting muscle. Their results were in accord with the collapse of water structure in contraction. Kaminer (1962) supported the theory by showing that contracted muscle tends to lose water, is partially dehydrated.

REFERENCES

Astbury, W. T., Beighton, E., and Weibull, C. (1955). *Symp. Soc. Exp. Biol.* 9, 282.

Bailey, K. (1948). *Biochem. J.* 43, 271.

Banga, I., and Szent-Györgyi, A. (1941-1942). *Stud. Inst. Med. Chem. Univ. Szeged* 1, 5.

Borbiro, M., and Szent-Györgyi, A. (1949). *Biol. Bull.* 96, 162.

Bowen, W. J., and Mandelkern, L. (1971). *Science* 173, 239.

Bradley, D. F., and Wolf, M. K. (1959). *Proc. Nat. Acad. Sci. U.S.* 45, 944.

Bratton, C. B., Hopkins, A. L., and Weinberg, J. W. (1965). *Science* 147, 738.

Dragomir, C. T. (1971). *J. Theor. Biol.* 31, 453.

Ebashi, S., and Endo, M. (1968). *Progr. Biophys. Mol. Biol.* 18, 125.

Edsall, J. T. (1930). *J. Biol. Chem.* 89, 289.

Engelhardt, W. A., and Ljubimova, M. N. (1939). *Nature (London)* 144, 669.

Huxley, A. F. (1957). *Progr. Biophys. Biophys. Chem.* 7, 255.

Huxley, H. E. (1963). *J. Mol. Biol.* 7, 281.

Huxley, H. E. (1969). *Science* 164, 1356.

Kaminer, B. (1962). *J. Gen. Physiol.* 46, 131.

Kühne, W. (1863). *Virchow's Arch. Pathol. Ana.* 26, 222.

Ling, G. N. (1962). "A Physical Theory of the Living State." Ginn (Blaisdell), Boston, Massachusetts.

Mihalyi, E., and Szent-Györgyi, A. G. (1953). *J. Biol. Chem.* 201, 189 and 211.

Meuller, H. (1966). *J. Biochem. (Tokyo)* 240, 3816.

Muralt, A. V., and Edsall, J. T. (1930). *J. Biol. Chem.* 89, 315 and 351.

Pfann, W. G. (1962). *Science* 135, 1101.

Pfann, W. G. (1967). *Sci. Amer.* 217, 62.

Rice, R. V. (1961). *Biochim. Biophys. Acta* 52, 602; 53, 29.

Rice, R. V., Brady, A. S., DePue, R. H., and Kelly, R. E. (1966). *Biochem. Z.* **345**, 37.

Slayter, H. S., and Lowey, S. (1967). *Proc. Nat. Acad. Sci, U.S.* **58**, 1611.

Straub, F. B. (1942). *Stud. Inst. Med. Chem. Univ. Szeged* **2**, 3.

Straub, F. B. (1943). *Stud. Inst. Med. Chem. Univ. Szeged* **3**, 23.

Szent-Györgyi, A. (1942). *Stud. Inst. Med. Chem. Univ. Szeged* **1**, 17.

Szent-Györgyi, A. (1945). *Acta Physiol Scand.* **9**, Suppl. 25.

Szent-Györgyi, A. (1947). "The Chemistry of Muscular Contraction." Academic Press, New York.

Szent-Györgyi, A. (1949). *Biol. Bull.* **96**, 140.

Szent-Györgyi, A., and Kaminer, B. (1963). *Proc. Nat. Acad. Sci. U.S.* **50**, 1053.

Szent-Györgyi, A. G. (1952). *Fed. Proc., Fed. Amer. Soc. Exp. Biol.* **11**, 1.

Szent-Györgyi, A. G. (1953). *Arch. Biochem. Biophys.* **42**, 305.

Weber, H. H. (1934). *Pfluegers Arch.* **235**, 205.

IV

BIOLOGICAL STABILITY AND THE EVOLUTIONARY PARADOX

The genetic pattern we inherit determines our entire makeup, as clearly shown by the similarity of identical twins. Professor Piccard the astronaut and deep-sea explorer and Professor Piccard the chemist were so alike that, the story goes, as students they could get two haircuts for the price of one: one of them having his hair cut and the other going to the same barbershop the next day, reproaching the barber for the poor job he had done and demanding a recut.

Alteration of the pattern entails grave consequences. A virus infection is essentially the introduction of a foreign pattern. Some time ago I put aside my lunchbox filled with food. When I opened it a few days later I found it filled with maggots, a new pattern having been introduced into the box by a fly. Life itself can be looked upon as the infection of our globe by a self-propagating pattern.

The human pattern is preserved chiefly in nucleic acid, and is transferred from generation to generation in DNA. Evolution is, essentially, a gradual change in this pattern that occurs through molecular accidents. There is no preconceived plan. Monod, in his classic book "Chance and Necessity" (1970), compares this process to playing roulette, where the number called is purely a matter of chance. The paradox of evolution is that this random process has led to the creation of such improbable structures as our own. It gives the impression that Nature knew from the start where she wanted to go, what she wanted to achieve; that she had the end, *telos*, in sight. Many scientists find themselves in a quandary. The idea that such improbable structures as human beings could have arisen from random accidents does not seem acceptable, while "teleology," the idea that all this has been planned in advance, has to be rejected. Teleology is said to resemble an attractive lady of doubtful repute whose company we cherish but in whose company we do not like to be seen. The geneticists' answer to the dilemma is that there was plenty of time for everything, but I do not think that an eternity would be long enough for building a Greek temple by randomly throwing bricks about.

The paradox can be resolved by accepting the supposition that a selective mechanism accompanies random mutation. Darwin was the first to propose a selective mechanism in the "struggle for life" and the "survival of the fittest." Monod alludes to an earlier selection, a decreased chance for reproduction. I feel that mutation and selection must be coupled still more intimately, both being basic attributes of life. For a clue to this problem I always searched for essential differences between the living and the inanimate. One such difference is known to everybody: inanimate machinery gets worn out by use, while live machinery improves by use and is impaired by inactivity. If you use your car too much it becomes worn out, while if you walk your legs become stronger.

The analogy between legs and cars may be fascinating but is useless experimentally. However, a century ago Bowditch, a young American working in Ludwig's laboratory (1871), described a simple experiment which may be a related phenomenon. What Bowditch did was to eliminate the normal apparatus of excitation in a frog's heart, making it dependent on electric stimulation. Then he stopped stimulation for a little while, probably expecting to find the first beats after the pause stronger, the heart having rested. What he found was the opposite: the first beat after the pause was weaker, the weaker the longer the pause. The second beat was somewhat better, each successive beat being somewhat stronger than the previous one, the peaks on his tracing rising as a staircase. This phenomenon, accordingly, is called "the staircase." What does this mean? It means that function generates function, motion generates motion, life generates life, while inactivity begets inactivity, makes motion and life fade away.

Some time ago Hajdu and I (1952; Hajdu, 1953) wondered about the chemistry of this phenomenon. We found that during the pause potassium was accumulated in the fibers; this is why the heart became weak. When the heart was allowed to beat, the potassium was pumped out again, order was restored.

Potassium and sodium have, in a normal heart, a very specific distribution: most of the potassium is inside and most of the sodium outside the fibers. During rest, this very specific distribution becomes mixed up and the potassium tends toward a random distribution. Possibly, other ions such as calcium behave similarly. If the heart does not work it loses its specific structure, tends to become a homogeneous soup. Expressed more scientifically, in rest the entropy increases, while in work it decreases. While we work we spend the negative entropy of our food, but some of it we retain, maintaining our own inner order. Thus in order to be able to maintain our pattern we have to work.

We recognize life by its actions; motion is one of them. *Quo cicior motus, eo magis motus*, the faster a motion, the more of a motion it is. The more life does, the more life it is; the more negative entropy is liberated, the more can be retained of it. Life supports life, function builds structure, and structure produces function. Once the function ceases the structure collapses; it maintains itself by working. A good working order is thus the most stable state. The better the working order the greater its stability and probability. In inanimate systems the most stable state is at the minimum of free energy and maximum of entropy. This is "physical stability." In living systems the opposite is true. The greatest stability is at the maximum of free energy and minimum of entropy, which corresponds to the best working order. This is "biological stability." In terms of biology, physical stability means death. Life tends toward biological stability which can be maintained only by function, and if there is no function the system slips toward physical stability. Accordingly, a favorable mutation which leads to a better working order will add to stability and will have a greater chance to be maintained, as compared to a senseless one which has no function. Even old, well-established processes fade if unused. This may be one of the factors driving the living organism toward perfection. Nature may not know where she is going but nevertheless improves herself. As to the mechanism of disintegration in inactivity a quote from Lewis and Randall's thermodynamics (1923) may give us a hint: "We often assume the existence of several equilibrium states toward which a system may tend, all these states being stable, but representing higher or lower degrees of stability." Inactivity may mean a gradual descent on this scale till final disintegration, while activity could entail a rise on the scale. Vitamins may be the substances of which there were plenty in our original surroundings, the tropical jungle. There being no need to make them, we forgot how to make them. The synthetic machinery disintegrated. In a car the screws and bolts gradually are loosened by use while in the living system they are tightened.

It is possible to draw up more concrete pictures of the atrophy of inactivity. If in Fig. 6b the globule of the H meromyosin is linked at every contraction to the troponin attached to the actin fiber at every 400 Å, then every contraction cycle will leave this globule at the right place. Should there be no contraction the globule may be carried away by random fluctuation and will be unable to find the troponin in excitation. Similarly, we can assume that in every contraction the K ions become bound at the right place on the contractile protein. In inactivity they will tend to wander away and contraction will be unable to bring them back. If we allow them to wander away too far this state may become irreversible and the whole system will collapse. In this light Katz's findings (1971) seem to be important. He found that even at rest there are little microcyles of contraction that do not integrate to a full macroscopic contraction. Such permanent functional activity may be important for maintaining the negative entropy of the living system, keeping things in their place, keeping the screws tightened. If Katz's mini-potentials and microcyles cease, as is the case when the nerve is cut, the whole system degenerates. This may explain how the nerve contributes to the maintenance of vitality, and why in a depressed state we age more rapidly.

All this may seem to be theoretical, but the distance which separates abstruse theory from the hospital bed is not as great as generally believed. Hajdu and I found that digitalis drugs, which have contributed so much to the relief of human suffering by improving the function of sick hearts, do so by decreasing entropy, helping the heart maintain its specific structure and ionic distribu-tion. In a way digitalis makes the heart more alive. Doc-tors may eschew digitalis, being afraid to "drive" a sick heart. Digitalis does not "drive" the heart; it only en-ables it to do its work better.

If mutations which create a new function or improve an existing one have a greater chance to be maintained than senseless ones, then the body will gradually create the machinery for which there is a useful function. Perhaps I could characterize the situation by saying that man is not able to talk because he happened to develop a speech center, but he developed his speech center because he had something to say.

It would be erroneous to look upon the deterioration of unused structures and functions as something purely negative. I think it is a positive building principle, an attribute of life which helps to adapt and develop new machinery. Adaptation and development involve not only the creation of something new, but also the discarding of what has become useless. Occasionally this may also lead to trouble. If cell cultures are intermittently deprived of oxygen they adapt to the new situation and shunt back to the anaerobic proliferative state, discarding the mechanism of biological oxidation. If this change becomes "constitutive," the cells will be unable to return to the aerobic way of living and cancer develops.

REFERENCES

Monod, J. (1970). "Le hasard et la nécessité." Seuil, Paris.

Bowditch, H. P. (1871). *Ludwigs Arb.* **6**, 759.

Hajdu, S. (1953). *Amer. J. Physiol.* **174**, 371.

Hajdu, S., and Szent-Györgyi, A. (1952). *Amer. J. Physiol.* **168**, 171.

Katz, B. (1971). *Science* **173**, 123.

Lewis, G. N., and Randall, M. (1923). "Thermodynamics," p. 19. McGraw-Hill, New York.

V

CELLULAR ARCHEOLOGY AND BIOENERGETICS

1. THE FUEL OF LIFE

Three things are needed to build and maintain a living system: material, information (pattern), and energy. It is generally believed that the "primordial broth" in which life originated contained organic molecules, one part of which could serve as building material, while the other could serve as foodstuff, as a source of energy. There is no life without energy, and the question was how the living system could release energy from its foodstuff molecules?

A molecule is, essentially, a cloud of electrons, held together by nuclei, and so its energy can be no other

44

than electronic energy. Strictly speaking, an electron, in itself, has no energy, as water has no energy. Water can give off energy and drive machines when it drops from a higher to a lower level, say from the top to the bottom of Niagara Falls. Similarly, the electron can give off energy and drive the living machine by dropping from a higher to a lower energy level. A high energy level means that the electron is held loosely (has a low ionization potential) and can drop to a lower energy level (of high ionization potential). At the lower level, it will be held more strongly, that is, more energy will be needed to lift it to a higher level or detach it from the molecule altogether. So our first question is how electrons can be made to drop from a higher to a lower energy level?

Mendeleef has ordered the elements into a single system. Tho two upper rows of his table are reproduced in Fig. 9. The first and last column contain the noble gases. The numbers are the atomic numbers which are equal to the number of electrons in the atom.

Atoms tend to resemble a noble gas, and have the same number of electrons as their nearest noble gas. To achieve this, the elements on the left side of the table would have to give off electrons while those on the right side of the table would have to take up electrons. Li, for instance, would have to give off one of its three electrons to have two, as He has, while F would have to take up one electron to have ten as Ne has. This make the elements on the left side positive, those on the right side negative. Elements on the left side will tend to donate

	H¹							
He²	Li³	Be⁴	B⁵	C⁶	N⁷	O⁸	F⁹	Ne¹⁰
Ne¹⁰	Na¹¹	Mg¹²	Aℓ¹³	Si¹⁴	P¹⁵	S¹⁶	Cℓ¹⁷	Ar¹⁸

Fig. 9. Upper rows of Mendeleef's table.

electrons while those on the right will tend to accept them.

The positive atoms of the left can be made to give off energy and have their electrons go to a lower energy level by interacting with a negative element from the right side, which will take up the electrons they tend to give off. For instance, H, from the left, will be able to give off its electron and release energy by interacting with O, an element from the right. In this interaction water will be produced and the electron of H will go from its high level to a lower level in H_2O, releasing the difference in energy. To put this into common language: by burning, oxidizing H, we can produce water and energy. At present, the energetics of the biosphere is based on this one reaction, the interaction of H and O, the separation and reunion of the two. The energy needed for the separation is supplied by light (photosynthesis), while the energy released in the reunion drives life (biological oxidation).

A glance at the atomic table shows the exceptional nature of H. It forms a group by itself. No other element has similar properties. Being the smallest, it is also the most mobile of atoms. It consists of an electron and a proton. A proton could not hold two electrons, but could easily part with its own. So we can say that, in a way, H is a package of energy which only has to interact with an atom (or atomic group) from the right side of the table to release its energy. Attaching H to a molecule means lending energy to it. Since in H the electron and proton are loosely coupled, by attaching an H we essentially attach an electron. So the chemists look upon both processes, attaching an H or attaching an electron, as more or less identical processes, and call it "reduction." Detaching oxygen is also a "reduction." The opposite of reduction is "oxidation," which means either detaching H, or electrons, or adding oxygen. By reducing a molecule we increase its energy, by oxidizing it we decrease it.

The H atom occupies a very special place for another reason. It consists of an electron and a proton. A proton is an H ion, H^+, and H ions are constituents of water, the universal biological solvent, a small part of which always dissociates:

$$H-O-H \rightleftharpoons H^+ + OH^-$$

This "dissociation of water" has a constant of 10^{-14}. A small part of the water molecule is always present in this dissociated state. For pure water the concentration of protons is 10^{-7} M. So if a molecule gives off a proton, this proton simply merges with water, if a molecule, say a sulfer atom, acquires an electron and becomes S^-, this S^- can easily pick up a proton from the water and become SH.

That detaching H and attaching O are equivalent processes, called "burning" or "oxidation," is easy to see. If we would detach from an organic molecule, say CH_4, H atoms pairwise and replace them with water, H–OH, then at every step the molecule would become richer in O and poorer in H, till it would become $C-(OH)_4$. Then if we would detach two water molecules, H–OH, from it, we would be left with CO_2 and the two molecules of water. In other words, we have oxidized, burned, the molecule completely. All we would need for this process would be activators and "H–acceptors," that is substances which could take up the detached H's.

The carbon atom is in the middle of the Mendeleef table. Thus it is neither positive nor negative and can unite with other C's to form long chains or rings, and can attach both positive and negative atoms. A foodstuff molecule is essentially a carbon structure with H's fixed to it, similar to hats on a hatrack. The living systems take these H's off and replace them with water. The foostuff is essentially an H donor while O_2 is an "H acceptor." *H is the fuel of life.*

When life originated, there was no O_2 in the atmosphere, so the first primitive living systems had to use some other H acceptor, and their metabolism had to consist of simply taking H's from foodstuff, liberating and using their energy by attaching them to the acceptor they could find. The only complication was that one can feed only when there is food, while life demands energy all the time. So the energy has to be stored. We solve this problem by storing food in the form of fat or carbohydrates. The first primitive living structures could not afford such a complex method and simply had to store the H's or electrons as such.

As a rule, in foodstuffs, the H's are strongly bound to the molecule, so in order to liberate them their binding has to be loosened, "activated" (the "activation energy" of their bonds has to be decreased). Probably, the first living macromolecules or macromolecular complexes had such "dehydrogenase" activity. Today this loosening or activation of H is accomplished by specific enzymes, the "dehydrogenases," or "H–activating enzymes." Dehydrogenation, having been one of the earliest biological processes, is still the foundation of our metabolism, and the dehydrogenases belong to our most important enzymes. The energy household of the first living system had to be very simple, consisting of taking H atoms or electrons from the foodstuff molecules, storing them for use according to need, and keeping them loosely bound. These primitive organisms must have had many things in common with present-day anerobic bacteria.

The structure of the energy-producing metabolic apparatus reflects the history of life. When life originated there was no light and no oxygen, the light being absorbed by a heavy layer of water vapor. In this first anaerobic period, life energy production must have been based chiefly on the possibility of taking H's or electrons from foodstuff, while the continuity of life was safeguarded by the ability of unlimited proliferation. Fermentation, essentially, consists of taking off and attaching H, reshuffling structure. It is linked to unlimited proliferation and anaerobic energy production.

As the earth cooled and the water vapor condensed, long wavelength red light could get through. To catch it, life developed a green dye, chlorophyll, green meaning that red is absorbed. The energy of these photons sufficed to separate the H from the O of water. This invested energy could be regained by reversing the process, attaching H to O, that is using O as the final electron acceptor and "burning" the H. Probably, the energy of the H was, at first, directly converted into the energy of ~P's (Arnon's cyclic phosphorylation) (Arnon, 1971; Arnon et al., 1967). Later, a bulkier and more efficient and complex oxidative machinery was developed which, being the last addition to the energy-producing mechanisms, is also most easily dispensed with. So when the cell divides it disassembles, partly or wholly, this oxidative mechanism and reverts to the simple and older anaerobic methods of H pooling and fermentation as has been shown by the classic work of Warburg (1966). Cell division and anaerobiosis are coupled.

2. THE H POOL

In 1922, Hopkins, working with Dixon, made observations which he was unable to interpret, though he had an almost uncanny insight into Nature's ways. His experiments, although simple, usually turned out to be of fundamental importance. He knew intuitively what was essential. His earlier work showed that in the cell there were two kinds of sulfhydryl groups, SH "fixed" and SH "soluble." The fixed ones were bound to the cellular structure while the soluble ones were part of a polypeptide, gluthathione, which he discovered. SH is easy to detect with nitroprusside with which, at a high salt concentration and pH, it gives an intensely colored purple compound, probably a charge-transfer complex, the absorption of which shows a sharp maximum at 515 mμ. When oxidized, the SH's lose their H and the S's join to form S–S, disulfide. The sulfhydril groups, SH's, have peculiar properties. On the whole they are not very reactive, but one S reacts readily with another S, and the H of the one readily shifts to the other. If an excess of oxidized gluthathione, GS–SG, is added to a protein which has fixed SH's, part of the GS–SG becomes reduced to two HSG's, while the fixed SH becomes oxidized to a great extent to disulfide, the reaction coming to equilibrium when most of the fixed SH is oxidized to S–S. In many ways the S and SH show unusual behavior. Huggins, Tapley, and Jensen (1951) found, twenty years ago, that the H, in protein solutions, can "jump" from one S to another, while Gordy and his associates (1955, Gordy and Miyabawa, 1960) showed that electrons placed on a protein become localized on an S. The negatively charged S⁻, then, may capture a proton to form SH.

Hopkins and Dixon suspended muscle in a solution of GS–SG. The GS–SG oxidized the fixed SH to SS while a corresponding amount of the GS–SG was reduced to SHG. The reaction came to a halt when most of the fixed SH was oxidized. At this point, evidently, equilibrium was reached. The quantity of SHG formed could be estimated by allowing it to reoxidize in air to GS–SG and measuring the O_2 absorbed. If instead of O_2 methylene blue was added, it acted as H accepter, and became reduced to leucomethylene blue. All this might have been expected. What Hopkins and Dixon probably did not expect was that muscle reacted the same way after having been boiled with water, extracted with hot alcohol, dried, and powered. They must have been even more suprised when they found that muscle treated with excess GS–SG reduced the same amount of GS–SG again after treatment with a fresh GS–SG solution. It did the same over and over again even if resuspended ten times. Always a new amount of H leaked out of it, reducing GS–SG. If the muscle powder was suspended in GS–SG in the presence of oxygen, the SHG formed was reoxidized continuously, and a continuous O_2 uptake resulted. The muscle seemed to "respire." At the beginning of the experiment CO_2 was produced with a respiratory quotient of one, but the quantity of CO_2 formed rapidly fell approaching zero asymptotically though the leaking out of H continued. This suggests that the initial CO_2 production was accidental and that the essential reaction was the giving off of H. Thus there had to be an H pool. If the muscle, treated exhaustively with GS–SG, was treated with cysteine or thioglycolic acid, the pool filled up again and continued to reduce GS–SG. The pool could thus be emptied and filled up. All this fitted nowhere into the then existing knowledge nor does it fit into existing knowledge today. The function of this pool did not involve heat-sensitive enzymes.

These experiments seemed most fascinating to me, and I wanted to repeat them with different materials and methods. Fresh muscle is not an ideal material, for it is difficult to cut it into equal, thin slices. Nor was glutathione ideal. Its nitroprusside complex is unstable, and measuring the quantity of SH by means of oxygen takeup on autoxidation is slow and cumbersome.

My pet material is the mouse. It cannot be cut into equal slices either, but its small intestine is a thin tube, more than 40 cm long, which, if cut open, forms a thin tape that can be cut into equal segments.*

In my search for a substitute for glutathione I reasoned that if the SH pool represents the reserve energy which has to feed the cell during its division, then division must be facilitated by substances which facilitate the leaking out of the H from the pool. I was thus looking for a substance which would promote cell division. Sollman, in his textbook on pharmacology (1957), states that thiourea is used in medicine to promote the granulation of torpid wounds. Rachmilewitz, Rosin, and Doljanski (1950) discovered that thiourea, administered per os to rats in large doses, produces an outburst of cell division in the liver. Thiourea is an equilibrium mixture of two forms (Fig. 10) with the equilibrium far to the left. It can be reduced to SH-thiourea by adding two H atoms (Fig. 11). The quantity of the reduced thiourea can easily be estimated because, containing SH at high ionic concentration and pH, it forms a highly colored complex with sodium nitroprusside [sodium nitroferricyanide $Na_2Fe(CN)_5NO \cdot 2(H_2O)$], which has a sharp absorption maximum at 515 mμ that can be used for its quantitative estimation. SH-glutathione also forms a similar compound with nitroprusside, but this compound is very unstable. In two minutes 30–40% of it decomposes. The thiourea complex is much more

---

*The small intestine consists of two parts. The smaller half, closer to the stomach, is thicker than the lower part. A 1-cm piece of the proximal part weighs about 30 mg/cm, while the lower, thinner part weighs about half as much.

$$\begin{array}{ccc} NH_2 & & NH \\ | & & | \\ C=S & \rightleftharpoons & C-SH \\ | & & | \\ NH_2 & & NH_2 \end{array}$$

Fig. 10. Thiourea.

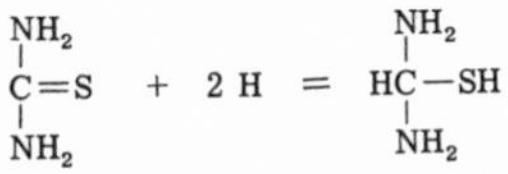

$$\begin{array}{cccc} NH_2 & & & NH_2 \\ | & & & | \\ C=S & + \ 2 \ H & = & HC-SH \\ | & & & | \\ NH_2 & & & NH_2 \end{array}$$

Fig. 11. Reduction of thiourea.

stable. Its color diminishes in two minutes by only 2–3%.

Thiourea is readily reduced by SH-glutathione,* so it can be expected to be reduced also by the "fixed" SH which reduces glutathione. The reduction of thiourea is thus analogous to the reduction of GS–SG in Hopkin's experiments (1925). My experiment was simple. I excised the small intestine of a mouse, cut it open, rinsed it in saline, then cut it into equal 2–2.5 cm pieces. I then immersed these pieces in 2 ml saturated ammonium sulfate, added 0.2 ml 10% thiourea, 0.2 ml 10% nitroprusside, and 0.1 ml ammonia. On incubation the fluid gradually turned purple, reaching in twenty minutes a very intense color, indicating the formation of reduced thiourea. In the control experiment no gut was added. There was a relatively slow and weak reduction in the control for which correction had to be made. The quantities of thiourea reduced are of the same order as

*That SH-glutathione readily reduces thiourea can easily be demonstrated by mixing a solution of SH-glutathione with a solution of thiourea. The nitroprusside reaction of pure SH-glutathione rapidly fades while the solution of pure thiourea shows no color with nitroprusside at all. The mixture of the two shows the long-lasting nitroprusside reactions of reduced thiourea.

the quantities of GS–SG reduced in Hopkins' experiments, though somewhat (2–3 times) higher per gram of tissue.

In these experiments the system could not come to equilibrium since the product, the reduced thiourea, was bound by the nitroprusside. Figure 12 shows the result of an experiment; 0.8 absorbance corresponds to the reduction of 0.05 mg thiourea.

3. BIOLOGICAL RELATIONS

The experiments of Hopkins and Dixon with glutathione, as well as my experiments with thiourea, indicate that the quantity of H or electrons stored by tissue is considerable. This brings to light a new quality and function of protein: it acts as a battery, as an accumulator. The anaerobic cells seem to live on this H or electron pool charged by the H's or electrons taken from the foodstuff by means of its H–activating enzymes. The function of the battery itself, the storage and release of energy, does not involve enzymes, which are defined as substances inactivated by heat, extreme pH's, or anhydrous solvents. Hopkins worked with boiled muscle extracted with alcohol; I worked at an extreme pH and salt concentration.

My own experience is at variance only at one point with that of Hopkins and Dixon who found that heat does not destroy the pool. This, in my experience, is true only to a limited degree. I found that short heating to 100°C reduced the activity of the pool in the gut by about 50%, while more prolonged heating at pH 4 completely destroyed it. This indicates that conformation and coordination of protein plays an important part in this function. We are thus confronted with two processes. One is the storage of H's or electrons, the other is their mobilization over the fixed SH.

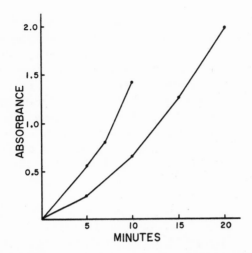

Fig. 12. Reduction of thiourea by the small intestine of the mouse. The average weight of the single pieces was 32 mg. They were suspended in 2 ml saturated ammonium sulfate to which 0.2 ml 10% thiourea and 0.2 ml 10% nitroprusside solution plus 0.1 ml strong ammonia were added. Absorbance was measured at 515 $m\mu$. The curve on the left side of the figure was obtained with gut incubated for 30 minutes in 1% hydrazone solution at pH 7. Temperature, 38° C.

As to storage the most likely assumption seems to be, at present, that the electrons of the pool are the lone electrons of the N's of the peptide bonds. Birks and Slifkin (1963; Slifkin, 1964) have shown that the amide N can act as electron donor in charge-transfer reactions. The same has been shown in my laboratory by McLaughlin (1968), as well as by Kimura and myself (1969). The storage, itself, does not involve specific configuration or coordination. Possibly, the two isomeric states of the peptide bond are involved in these reactions. What was interfered with by boiling had to be the mobility of the electrons: in boiled tissue they could no longer reach the SH. Peptide links can become conjugated by means of H bonds and may thus form a continuous network within the protein through which the electrons can move and reach the fixed SH. This network has to be destroyed by heat. This could also explain the difference in the behavior of proteins which are involved in cellular energetics and the behavior of proteins which have no energetic function such as serum albumin. All proteins or amide groups can equally serve as electron donors toward diffusible acceptors such as chloranil (used by Birks and Slifkin) or N–methylphenazonium sulfate (used in my laboratory). Proteins, such as serum albumin, which have an energetically passive role, need not build such a conjugated system which makes one single electronic system out of the whole protein molecule.

During early anaerobic life, the H pool may have been constantly on tap, so that the cell could divide whenever there was sufficient food to fill the pool. There had to be no limit to proliferation. The situation changed when O_2 appeared and the energy-rich biological oxidation had become the main energy source, as pointed out by Warburg. This new wealth of energy opened the way to differentiation and to the building of complex multicellular organisms. Proliferation now had to be suppressed in the interest of harmony of the whole organism, which could be done by turning off the tap of the H pool, that is, inactivating reversibly the SH and binding the H's or electrons. This binding could be done by electron acceptors formed in biological oxidation, while the SH could be inactivated reversibly by carbonyl's forming hemimercaptals with it (Fig. 13). The ready formation of hemimercaptals and SH was demonstrated by Schubert (1935, 1936) twenty-five years ago.

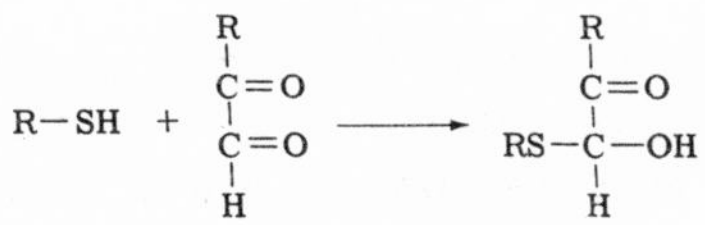

$$R-SH \; + \; \begin{array}{c} R \\ | \\ C=O \\ | \\ C=O \\ | \\ H \end{array} \longrightarrow \begin{array}{c} R \\ | \\ C=O \\ | \\ RS-C-OH \\ | \\ H \end{array}$$

Fig. 13. Formation of a hemimercaptal.

When the cell divides it has to break down its bulky oxidative mechanism and revert to the more archaic fermentative method of energy production and the use of the H pool. Proliferation and anaerobiosis, as well as rest and oxidation, are coupled. The H pool contains enough H to cover the energy needs of the cell during one division, to be replenished later in interphase. In rapidly growing tissue there is no time to replenish the H pool, so we can expect to find it depleted. The experiment showed that embryonic tissue and malignant tissue (Sarcoma 180 and Krebs 2) are practically unable to reduce thiourea, contrary to other tissues which reduce thiourea in the following order:

liver > intestine > kidney > heart > lung > spleen

There is an unsolved puzzle about fixed SH: its "masking." As is generally known, tissues, in the native state, show a poor SH reaction with nitroprusside. If "denatured" by heating, the SH reaction greatly increased. So the SH, in native material, is "masked" and unmasked by denaturation, be it by heat or trichloroacetic acid. What masking really means, nobody knows. It was tacitly assumed by some that the SH is, somehow, located deeply in the protein structure and so is not accessible to the SH reagent. Klotz (1962) assumes that the solid water sheet covering it may interfere with the diffusion of the reagent. That the mechanism of masking may be different is indicated by the fact that alcohol does not unmask the SH, though it denatures the protein and destroys water structures. It seems likely to me that the masking is due to the formation of a complex, probably a hemimercaptal, which is formed when the H pool has to be shut off. Hemimercaptals are not very stable compounds and can be split by heat or by strong acid such as trichloroacetic acid, but not by alcohol. Thus the masking may be an expression of regulation. If this assumption is correct than incubation in the presence of hydrazine should free the SH and increase the rate of reduction of thiourea. It actually does so. If the gut is incubated for thirty minutes with 1% hydrazine (pH 7), the rate of reduction of thiourea is found to be increased considerably (Fig. 12).

Nitroprusside, at a high ionic strength and pH, also forms a complex with SH and so can be expected to interfere with the reduction of thiourea. It actually does so. When the gut was suspended in saturated ammonium sulfate solution and only a small amount of nitroprusside, e.g., 0.1 ml of a 10% solution, was present, the reduction of thiourea began without delay. When the quantity of nitroprusside was doubled there was a delay of five–ten minutes. When the quantity was doubled once more it took fifteen minutes for the gut to break through the inhibition.

Yeast, suspended in ammonium sulfate, rapidly reduced thiourea, as indicated by the purple color developed in the presence of nitroprusside and ammonia. However, if starved, that is, if it is incubated in saline without nutrients for a few hours, it reduces no more. Reduction can be restored by suspending it for a short while in a nutrient solution allowing the H pool to be filled up. This shows its H pool to be an active constituent of the living system, closely linked to metabolic activity.

REFERENCES

Arnon, D. I. (1971). *Proc. Nat. Acad. Sci. U.S.* **68**, 2882.

Arnon, D. I., Stujimoto, H. Y., and B. D. McSwain (1967). *Nature (London)* **214**, 562.

Birks, J. B., and Slifkin, M. A. (1963). *Nature (London)* **197**, 42.

Gordy, W., and Miyagawa, I. (1960). *Radiat. Res.* **12**, 211.

Gordy, W., Ard, W. B., and Shields, H. (1955). *Proc. Nat. Acad. Sci. U.S.* **41**, 996.

Hopkins, F. G. (1925). *Biochem. J.* **19**, 787.

Hopkins, F. G., and Dixon, M. (1922). *J. Biol. Chem.* **54**, 529.

Huggins, C. B., Tapley, D. T., and Jensen, E. V. (1951). *Nature (London)* **167**, 592.

Kimura, J. E., Szent-Györgyi, A. (1969). *Proc. Nat. Acad. Sci. U.S.* **39**, 286.

Klotz, I. M. (1962). *In* "Horizons in Biochemistry" (M. Kusha and B. Pullman, eds.), p. 523. Academic Press, New York.

McLaughlin, J. A. (1968). *Proc. Nat. Acad. Sci. U.S.* **60**, 1418.

Rachmilewitz, M., Rosin, A., and Doljanski, L. (1950). *Amer. J. Pathol.* **26**, 195 and 937.

Schubert, M. P. (1935). *J. Biol. Chem.* **111**, 671.

Schubert, M. P. (1936). *J. Biol. Chem.* **114**, 341.

Slifkin, M. A. (1964). *Spectrochim. Acta* **20**, 1543.

Sollmann, T. (1957). "Manual of Pharmacology," 8th ed. Saunders, Philadelphia, Pennsylvania.

Warburg, O. (1966). "The Prime Cause of and Prevention of Cancer." Lecture at the meeting of Nobel Laureates, June 30, Lindau. (English ed. by Dean Burk, K. Triltsch, Wurzburg.)

VI

VEGETABLE DEFENSE SYSTEMS

1. PRINCIPLES OF DEFENSE

If we expose ourselves unprotected to sunshine we get a sunburn; our skin becomes damaged. The damage activates the tyrosinase system; the tyrosinase oxidizes tyrosine into a black pigment which then protects us against the sunshine. This cycle of events reflects the basic principle of defense: the cell contains an enzymic system which becomes activated by the damage and then protects against the damage. We also encounter the same principle in the case of mechanical damage where three different protective systems may be activated. One protects against bleeding. An enzyme is activated which produces fibrin and fibrin plugs the damaged blood

60

vessels. If the damage is irreparable, a proteolytic enzyme, cathepsin, is activated which eliminates the damaged part by digesting it. A third system activated is that of proliferation which heals the wound.

We know little about how these enzymic systems are kept in an inactive state prior to damage. There are various possibilities. Enzyme and substrate may be kept separated, the one being bound or enshrined in little vesicles that are broken down by damage. This method of inactivation is used by Nature not only in defense mechanisms but also in normal function. The proteolytic enzymes of the pregnant uterus, for instance, are kept enclosed in lysosomes which are stabilized by progesterone (Goodall, 1965, 1966). After delivery, the secretion of progesterone ceases, and the enzyme is released dismantling the hypertrophic muscle tissue. Acetylcholine is, in the same way, kept separated from the membrane in motor nerve endings.

Another method of keeping an enzymic system inactive consists of harboring the enzyme itself in an inactive form, as a proenzyme. In this case the damage will have to release the activator, so it will be the proenzyme and its activator that have to be kept separated. A classic example is prothrombin, which is activated by thrombokinase.

Keeping two substances separated within the narrow confines of the cell demands a very high degree of order. The damage thus, essentially, means disorder, an increase in entropy, one of the most basic parameters in Nature.

2. PLANTS AND ANIMALS

There is but one life on earth. We are all but recent leaves of the same old tree of life. There is no basic difference between the grass and the man who mows it. Life has differentiated to a great extent.

There are many major differences between plants and animals. One is that plant cells have a solid cellulose membrane. While this membrane is intact, the cell is inaccessible to bacteria and needs no protection against them. Plants have no blood circulation, but animals have; the latter can thus mobilize the defense of the whole body against a local lesion, while plants have to deal with it locally. This makes it difficult for plants to repair damaged cells. The damaged cells thus have to be eliminated altogether to prevent serving as food for bacteria.

3. DEFENSE AND REGULATION

In both plants and animals regulation of growth plays a major role in defense. Plants defend themselves against bacteria by regulating their proliferation, arresting it irreversibly, thus killing them. The borderline between defense and regulation is rather hazy.

Proliferation is a basic attribute of life involving all parts of the cell, and can best be regulated by changing one of its basic parameters. A dividing cell needs energy, so proliferation can be arrested by withholding energy. As has been discussed in a previous chapter, energy means H or active electrons, i.e., H or electron donors; it means a reducing atmosphere. So growth can be prevented by an oxidizing atmosphere by H or electron acceptors that will absorb active H atoms or electrons. Growth can thus be arrested by changing the D/A (Donor/Acceptor) quotient in favor of A. The stronger the acceptor the more powerful its action will be. Having no blood circulation which would dissipate poisons, plants can use rather powerful A's while animals must be more careful in their choice, and must have their A's attuned carefully to their needs.

There is but one acceptor which the cell can readily produce: the double bond. One of the two bonds of a double bond is a σ bond, the same kind of bond which holds most organic molecules together. The other is a π bond. Each double bond has two π orbitals, usually only one of which is occupied by electrons, the one with the lower energy. The other, the "antibonding" one, is empty and can accommodate an extra electron, serving as acceptor.

A small electronic system, such as a double bond, is not apt to accept an unbalanced extra electron which upsets its electroneutrality. This makes it a very poor acceptor. We can extend the system and make it a better acceptor by adding a second double bond to it in the $\alpha-\beta$ position. The two double bonds, being separated by one saturated C–C bond only, are "conjugated," and their π systems fuse. The more extensive such a system is, the more conjugated double bonds we add to it, the stronger an acceptor it will become.

There are various double bonds: C=C, C=O, or C=N. Each has its own peculiarities. The C=C, as a rule, is not very reactive. If activated, it is capable of irreversible additions, which makes it poisonous and unfit for a catalytic role. The most common acceptor is the C=O which can readily be formed, C and O being ubiquitous. If the C=O is located at the end of a carbon chain, the resulting substance is an aldehyde (Fig. 14A,a). If located inside the chain the resulting substance is a ketone (Fig. 14A,b). Ketones are thermodynamically satisfactory as acceptors but are not very reactive. Aldehydes are more reactive but thermodynamically less favorable, being hydrated (Fig. 14B). A ketonic and an

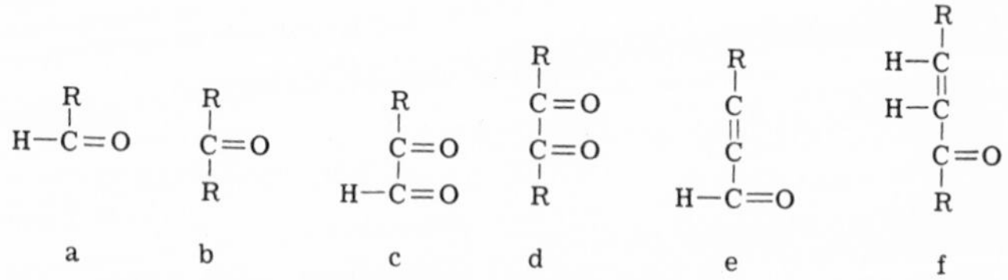

Fig. 14A. a, Aldehyde; b, ketone; c, aldoketone; d, diketone; e, unsaturated aldehyde; f, unsaturated ketone.

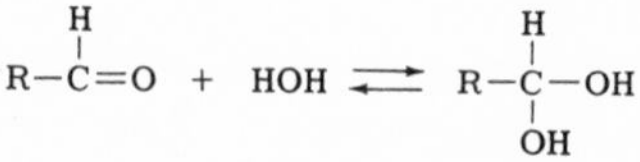

Fig. 14B. Hydration of an aldehyde.

aldehydic C=O make an aldoketone (Fig. 14A,c); two ketone C=O's make a diketone. Conjugated C=N bonds, as found in flavine and pteridine, are excellent acceptors. By having accepted an electron they become good donors, which makes them "electron transmitters."

The most active acceptors can be formed by placing two C=O's side by side on an aromatic ring which, in itself, is a system of conjugated double bonds. The resulting substance is an o-diquinone (Fig. 15a). The p-diquinone (Fig. 15b) is somewhat less active, the two O's being separated, though still connected by the conjugated bonds of the aromatic structure. These quinoid acceptors are most powerful, and Nature uses them chiefly with the intent to kill.

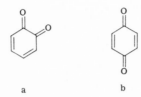

Fig. 15. a, o-Diquinone; b, p-diquinone.

4. THE POLYPHENOL OXIDASE SYSTEM

Fruits and vegetables can be divided, grossly, into two groups: those which discolor, become brown or black on damage, and those which do not. If you drop a banana, apple, or pear, the next day you will find a brown patch where the fruit was damaged. If you peel potatoes, or work with mushrooms, your hands get black. Lemons, oranges, watercress, and cabbage never discolor. The members of the first group are characterized by the presence of a polyphenol and a polyphenol oxidase. The plants which show no such discoloration are characterized by the absence of a polyphenol oxidase and the presence of an active peroxidase. Members of the first group are called *polyphenol oxidase plants*, while members of the second are called *peroxidase plants*.

Attracted by color, my first biochemical paper was concerned with the mechanism of the discoloration of polyphenol oxidase plants. It was known that this discoloration was due to the oxidation of a polyphenol (an aromatic substance with more than one phenolic OH group) to a polyquinone by a polyphenol oxidase (Fig. 16). Some of these ferments can even oxidize mono-

$$\text{(catechol)} + \tfrac{1}{2}O_2 \longrightarrow \text{(o-quinone)} + H_2O$$

Fig. 16. Oxidation of catechol by a polyphenol oxidase.

phenols, introducing a second OH group and then oxidizing the diphenol to a diquinone (Szent-Györgyi, 1925).

If the cut surface of these plants is treated with an alcoholic solution of guaiac resin an intense green color develops. Benzidine, under similar conditions, develops a brilliant blue color, both substances being oxidized to their highly colored quinoid form. Since these reagents give the same highly colored oxidation products if treated with peroxidase and peroxide, it was thought that the oxidation of polyphenols by the polyphenol oxidase involved peroxidase and peroxide. I could show that what happened was simply that the catechol derivative present was oxidized to an o-diquinone; the o-diquinone, being an excessively strong oxidizing agent, oxidized guaiac or benzidine to their colored oxides, and no peroxidase was involved.

Why do the polyphenol oxidase plants discolor only on damage? Polyphenol oxidase is a very powerful enzyme. How powerful it is can be demonstrated by allowing the crushed plant to act on a polyphenol (say catechol) in the presence of ascorbic acid. The ascorbic acid reduces any quinone formed immediately so that the oxidase can act unceasingly on the unoxidized substrate. The reducing power of the plant is relatively poor so that its tissues could never keep in step with the oxidase and reduce the quinone formed if the oxidase were active. Evidently, in the intact plant, the enzyme is inactive, and becomes activated only by damage. What, then, is the biological function of the enzyme? Quinones are too powerful an oxidizing agent to take part in normal respiration. Evidently the quinones are meant to destroy the bacteria which invade the damaged cells. How the enzyme is kept inactive prior to damage, we do not know. All we know is that damage activates it, any damage, be it mechanical or chemical. Chloroform is a fairly inactive agent, but dip a banana into chloroform for an instant, and the next day you find it has turned black. The quinones do more than kill bacteria. They tan the protein in fruits and vegetables, making them unfit food for bacteria. The tanned protein forms a protective sheet over the damaged tissue. The dark color seen in the damaged plant is not due to the quinones themselves, which are pale yellow, but to the protein complexes of the quinones. This binding of the quinone by the protein also prevents it from diffusing away and damaging neighboring cells.

This enzymic system has a high survival value. It is a most ingenious trap set for bacteria which penetrate the cell, hoping for a good feed: by damaging the cells they activate the enzyme and therefore commit suicide. The Creator must have had much fun thinking up this system.

The polyphenol oxidase system reflects the basic principles of defense. The damage elicits its own correction by activating an enzymic system which prevents proliferation by creating electron or H acceptors, shifting the D/A balance in favor of A.

The polyphenol oxidase not only protects plants, it also protected me. It was my first biochemical paper on polyphenol oxidase which attracted the attention of Sir Frederick Gowland Hopkins who invited me to his laboratory, just when my difficulties seemed insurmountable and I was about to give up scientific research.

There is no evidence for the participation of the phenol oxidase system in human physiology. Quinones are poisonous and incompatible with animal life. Nevertheless it would not be surprising if a related system were involved in animal physiology. The tyrosinase system which produces our pigments is closely related to it. Dopa is a nitrogen-containing polyphenol which also can be oxidized by polyphenol oxidase to a highly colored quinone, the oxidation power of which is abolished by an internal reaction with its own nitrogen. So dopa could be involved in animal metabolism. George C. Cotzias's work indicates an intimate relation between this substance and Parkinson's disease. The red color of certain nerve centers, such as the "nucleus ruber" (the "red nucleus") also suggests a phenol oxidase. Polyphenol oxidases are known to occur in insects.

5. THE PEROXIDASE SYSTEM

The "peroxidase plants" owe their name to the highly active peroxidase they contain. If their juice is pressed out, on addition of benzidine and hydrogen peroxide an intense blue color develops, indicating the presence of peroxidase which oxidizes the benzidine to its colored oxide. From the polyphenol oxidase plants I turned to the peroxidase plants. I pressed out their juices and added benzidine and peroxide. The expected blue color developed, but in some instances with a short delay of a fraction of a second, while a purified peroxidase developed the color instantaneously. This delay indicated the presence of a reducing agent which re-reduced the oxidized benzidine, so the color could develop only after the reducing agent was oxidized. I isolated the agent; it was an acid. I also found it in greater concentration in the adrenal cortex. Its elementary analysis and other properties indicated a derivative of an unknown carbohydrate.

Since "ignosco" means "I do not know," and "ose" means sugar, I called it "ignose." The editor of the *Biochemical Journal* reprimanded me for making jokes about science. "Godnose" was not any more successful. So, eventually, following a proposal made by the editor, it was called "hexuronic acid," which turned out to be a misnomer. I could obtain "hexuronic acid" in crystalline form on a small scale from various sources—cabbage, lemons, oranges—but preparations on a larger scale failed. I once invested all my spare pennies in a larger quantity of oranges; both the orange juice and my pennies went down the drain. I tried all the plants I could lay my hands on. The adrenal glands were the only material from which preparation on a larger scale succeeded, but they were not available in Europe in quantity. Professor E. C. Kendall of the Mayo Foundation of Rochester, Minnesota kindly invited me to his laboratory where the material from the nearby slaughterhouse of St. Paul could be used. I worked in Dr. Kendall's laboratory a full year, obtaining fifteen grams of crystalline material which I gave to Professor Haworth, a leading carbohydrate chemist, to analyze for details of chemical structure. He exhausted the supply without arriving at a conclusion (Szent-Györgyi, 1928).

At this point I went home to Hungary to help rebuild the scientific life destroyed by World War I. My town, Szeged, happened to be the center of the pepper ("paprika") industry, and pepper was the only plant I had never tried. At that time it was not cultivated in the United States. I have seen it once in Cambridge, England, but was told by the vendor that it was poisonous. One night, in Szeged, my wife gave me raw pepper for supper. I did not feel like eating it but did not have the courage to say so. It occurred to me that this was the only plant I had never tested, and so I said I would rather like to take it to the laboratory to see whether it

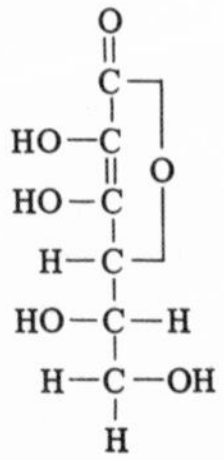

Fig. 17. Ascorbic acid.

contained "hexuronic acid." By midnight I knew that it was a treasure trove. A few weeks later I had one and a half kilograms of "hexuronic acid" in crystalline form which I distributed among scientists all over the world who wanted to work on it. This helped establish its structure (Fig. 17).

Scientists often have pet subjects and pet aversions. My pet aversions were vitamins and muscle. Vitamins were too glamorous, and the word "vitamin" conveyed no new basic knowledge. It means that we have to eat it to stay healthy. What we have to eat to stay healthy was, according to me, something for the cook to worry about. I considered right from the beginning the possibility that "hexuronic acid" was vitamin C—the unidentified vitamin that prevents scurvy—but I did not bother to find out if it was. Fate would have it that I was joined by a young biochemist of Hungarian origin, J. Svirbely. On my question "What can you do?" he told me that he could find out whether a substance was vitamin C. I gave him my "hexuronic acid," telling him that he would find it to be vitamin C. The experiment consists of feeding guinea pigs a vitamin C-free diet and adding the substance in question to the diet of some of the animals. One month later Svirbely reported that the animals, without supplement, stopped growing. With hexuronic acid they grew normally (Szent-Györgyi, 1930). Hexuronic acid was vitamin C, but to follow the classic rules we had to wait another month for the animals to die before publishing our results. With Professor Haworth we rebaptized "hexuronic acid" "ascorbic acid" (Haworth and Szent-Györgyi, 1933).

In addition to ascorbic acid I also found an "ascorbic acid oxidase" (1931), an enzyme which could oxidize ascorbic acid in the presence of air to dehydroascorbic acid (Fig. 18). The oxygen uptake was of the same order as the respiration of the undamaged plant (cabbage) which contained it, suggesting that ascorbic acid and its oxidase were actually involved in respiration. But no peroxide seemed to be formed and no phenolic substance was known to be present on which the peroxidase could act. So the story was incomplete and made no sense. A major link seemed to be missing.

The missing link may have been found later, the story beginning with my riding a horse in Jamaica fifteen years ago with a gentleman from Illinois. While I enjoyed my sumptuous breakfast he ate an odd-looking concoction, a mixture of yeast and wheat germ. He ate it, he said, because he used to have several grave colds every year, but since he had been eating this mixture he

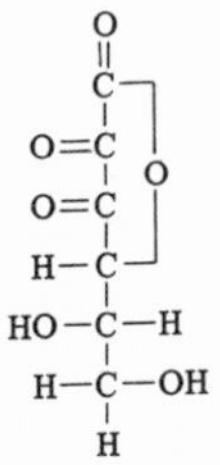

Fig. 18. Dehydroascorbic acid.

had had none. Having myself suffered from colds a great deal, I started to breakfast on wheat germ too, and since then I have had no colds either. Earlier, I had always been the first to pick up any cold, and had almost died twice of pneumonia. Now I look at other people with dripping noses and think "You should eat wheat germ and take ascorbic acid."

What could there be in wheat germ that does the trick? Some time ago it occurred to me that wheat germ is a "peroxidase plant." Bread, our staple food, is the grain of this grass. This grain contains at one end the pattern of life, the germ. Most of the important nutrients needed for life are located in this germ. The rest is chiefly starch. Man carefully mills out the germ before he eats the grain, rejecting what he needs for his health, while eating the rest which has only calories and no qualitative value. So if my missing link existed, it had to be found in high concentration in the germ; and perhaps this was the missing link which had kept my colds away.

This led to the idea of mixing yeast and wheat germ in the laboratory to see what happens, hoping for my hypothetical phenol to turn up. Soon I discovered that a Swiss chemist, L. Vouataz, had earlier made yeast act on wheat germ (1950). He found that if he eliminated the SH present by oxidizing it, a substance was formed which made the nitroprusside reaction of added SH-glutathione disappear irreversibly. Cosgrove and his associates (1952a,b) identified the substance as methoxybenzoquinone (Fig. 19a) associated with which they also found 2,6-dimethoxybenzoquinone (Fig. 19b). The latter did not interact with SH. Bungeberge de Jongh, Klaar, and Vliegenhart (1953) showed these substances to be present in wheat germ as glucosides. The role of the yeast thus seemed to be to split the glucosidic linkage and liberate the methoxybenzoquinone. Since

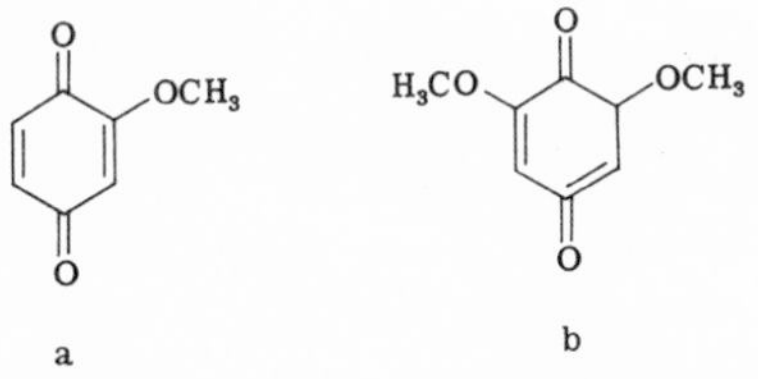

Fig. 19. a, Methoxyquinone; b, 2,6-dimethoxyquinone.

our body can do the same, I omitted the yeast from my breakfast.*

For a student of the peroxidase system these findings seemed most exciting. In the intact plant the methoxyquinone had to be present in reduced form as methoxyhydroquinone (Fig. 20a), "MH," or dimethoxyhydroquinone (Fig. 20b), "DMH." Could it be that this MH or DMH was the missing link? The methoxy group may have been introduced by the plant to lend specific properties to the molecule. Hydroquinone itself, without the methoxy group, easily makes additions to its double links, and methoxy groups were possibly also introduced to reduce this trend. Two methoxy groups eliminate addition completely. At this point Professor G. Fodor and his associate J. P. Sachetto came to my aid and synthesized all these substances for me. The "MH" was readily oxidized by peroxidase and peroxide to a vivid green-yellow quinone, so this could be the substance on which the peroxidase acted in the plant.

*My breakfast consists of the following: a sliced banana, over which is poured about two ounces of wheat germ, and then add milk. I finish with tea to which I add a heaping spoonful of a powder which I prepare by mixing 80 grams of ascorbic acid with 1 pound of confectioners sugar. The heaping spoonful contains about 1 gram ascorbic acid. The ascorbic acid lends a pleasant, slightly acid taste to the tea, replacing lemon. I repeat taking tea with ascorbic acid in the afternoon.

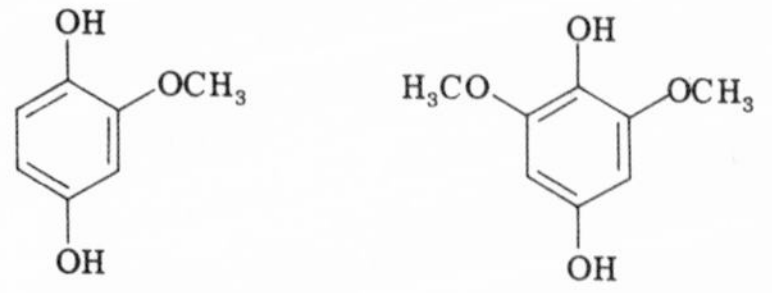

Fig. 20. a, Methoxyhydroquinone; b, 2,6-dimethoxyhydroquinone.

When MH was added to a plant juice, say the juice of cabbage, the ascorbic acid soon became oxidized and a deep purple color appeared; in all probability a semiquinone of MH formed. The plant juice catalyzed the oxidation of MH, but lost this catalytic activity on dialysis, to regain it when the ash of the plant juice was added. This meant that the catalyst had to be inorganic.

Plants contain manganese, Mn, in relatively high concentration. If the Mn was restored after dialysis the plant juice regained its catalytic activity. So it could have been Mn which catalyzed the oxidation of MH. Manganese does not catalyze the autoxidation of hydroquinone, so evidently the introduction of the methoxy group was responsible for this new reactivity. This change in catalytic activity initiated by the methoxy group seems rather specific. Reduced methoxytoluquinol behaved very similarly to MH, but only if the two methylmethoxy substitutions were in the 2,6 and not in the 2,5 positions. 2,6-Dimethoxyhydroquinone is autoxidizable, but its autoxidation is not promoted by Mn, whereas the 2,5-dimethoxyhydroquinone is not at all autoxidizable. This makes it likely that among the methoxyhydroquinones of wheat germ only the monomethoxy compound has a direct biological function.

What made these findings rather exciting was that H. Lundegaerdh showed about thirty years ago (1939) that manganese played an important role in vegetable respiration, which he was unable to explain. My experiments suggest that the role of Mn is to catalyze the autoxidation of the MH present. The Mn catalysis can easily be demonstrated by adding 0.01% MH to a 1.0% solution of p-phenylenediamine (Fig. 21a) in the presence of a pH 7 phosphate buffer. The oxidizing MH oxidizes the diamine to its purple diimine. This reaction is greatly speeded up by the addition of 0.01% manganese acetate, which, in itself, has no effect on the diamine. What made these findings still more exciting was that during the Mn-catalyzed oxidation of MH, peroxide, H_2O_2 , was formed. This added a second missing link to the peroxidase system. The peroxidase, which dominates the whole system, is inactive without peroxide, and nobody knew yet from where the peroxide could come. The formation of H_2O_2 could easily be demonstrated by adding o-phenylenediamine (Fig. 21b) to the mixture of peroxidase, Mn, and MH. The o-phenylenediamine has a somewhat more positive redox potential than the para compound, and is not oxidized by methoxybenzoquinone but can be oxidized by peroxide plus peroxidase, developing an intense color. A solution of o-phenylenediamine, containing Mn, developed this color on addition of peroxidase, but not without it. So evidently in the oxidation of MH, catalyzed by Mn, peroxide was formed.

The "peroxidase system" thus began to take shape, and its constituents could, tentatively, be put together to form a system. With manganese the MH or DMH could be oxidized to its corresponding quinone. The peroxide thus formed could interact with the peroxidase and oxidize a second molecule of MH or DMH to its

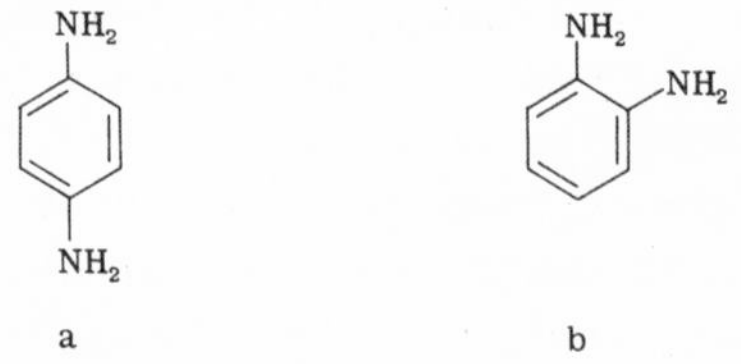

Fig. 21. a, p-Phenylenediamine; b, o-phenylenediamine.

quinone. The quinones were reduced by the ascorbic acid present, the ascorbic acid being oxidized to dehydroascorbic acid. The ascorbic acid oxidase may have contributed to this oxidation. The dehydroascorbic acid thus formed was re-reduced to ascorbic acid by the metabolic apparatus which involved the SH-glutathione, and dehydrogenases, establishing a reducing atmosphere with a high D/A ratio.

If a plant, say the cabbage leaf, is damaged mechanically, crushed, or frozen and thawed, the hydrogen-activating system becomes inoperative, and the dehydroascorbic acid, the oxidized MH, DMH, and the H_2O_2 are no longer reduced, and the strongly reducing atmosphere is turned into an oxidizing one which is unfit for bacterial growth. So here we find, essentially, the same principle as we found in polyphenol oxidase plants. In the latter the low D/A ratio was achieved by the production of strong acceptors. In the peroxidase plants it is achieved by decreasing the numerator of the ratio arresting H activation, arresting the production of D. In both cases, that of polyphenol oxidase and peroxidase plants, the primary event was the increase in entropy, the creation of disorder by the damage, followed by the consecutive change of an electronic atmosphere to a lower D/A, which made the system unfit for bacterial growth (Szent-Györgyi, 1970).

In plant juice MH and DMH are present in very low, catalytic concentrations, which makes it difficult to detect them; only juices of certain plants, such as watercress, produced by freezing and thawing, assume a purplish color on exposure to air, due in all probability to the oxidation of the MH present.

6. REMARKS ON HUMAN HEALTH

I always felt that not enough use was made of ascorbic acid. "If we lack ascorbic acid we develop scurvy. If we do not have scurvy, we have enough ascorbic acid." So the argument ran; the logic was impeccable. The flaw in this argument is that scurvy is not the first symptom of deficiency. It is a sign of the final collapse of the organism, a premortal syndrome, and there is a very wide gap between scurvy and a completely healthy condition. Good health is the state in which we feel best, work best, and have the greatest resistance to disease. Nobody knows how far we are from such a state. This could be established only by extensive statistical studies which are not available. Solutions to this problem are full of pitfalls. If, owing to inadequate food, you contract a cold and die of pneumonia your diagnosis will be pneumonia, not malnutrition, and chances are that your doctor will have treated you only for pneumonia.

Is our body such a poor mechanism that it has to break down every so often with a cold or other ailment? Or do we abuse our body, and feed it so poorly that its breakdown is comparable to the breakdown of an unlubricated engine? I am often shocked at the eating habits of people. What I find difficult to understand as a biologist is not why people become ill, but how they manage to stay alive at all. Our body must be a very wonderful instrument to withstand all our insults.

There is a widespread belief that expensive food is good food. This is very far from the truth. When I was a professor in Belgium, I learned that the second son of the king, the Prince of Liège, was in poor health, and had a temperature all the time. Belgian food is very good, and the king is very rich. Nobody could find out what was wrong with the ailing prince till somebody got the bright idea of giving him vitamin C, whereupon his condition cleared up. There was a similar case later in Sweden involving a diplomat's son. When I was in that country the last time, I was told that evidence had been found that vitamin C was completely harmless. An unbalanced individual had consumed incredible amounts of ascorbic acid without the least ill effect. Vitamin C is harmless. It does not injure your health nor your pocketbook, being a very cheap commodity.

Linus Pauling, one of the great scientists of our age, must be credited with having brought these problems into focus with his little book on vitamin C (1970). I concur with almost everything he says. I agree with his statement that individual needs for vitamin C vary within wide limits, and since you do not know what your needs are, you should take plenty of it before symptoms appear.

Symptoms of deficiency may show up in many ways. I know of one case involving a completely antisocial boy who was made congenial by the administration of vitamin C.

The arguments for a relatively low need of vitamin C are misleading. They are based partly on the fact that if we take larger doses of vitamin C, some of it is eliminated in the urine. What's the point of taking more of it if it is eliminated? This again may seem an impeccable argument. The situation is analogous to that of blood sugar. Blood contains 6 grams of sugar, so if we suddenly ingested 10 grams of it some would appear in the urine even though our daily need for sugar is of the order of a pound.

My finding wheat germ useful against colds is not at variance with Pauling's statements. If one is deficient in two factors, administration of one of them may help to some extent, while full benefit may be derived only by taking the two in combination. Should it be the combined effect of ascorbic acid and MH or DMH which protects me against colds, the situation in my lungs would be similar to that in the damaged peroxidase plant in which the combination of the two substances is needed to establish the oxidative atmosphere which protects against microorganisms.

I think I owe my 78 years to the combined effect of ascorbic acid and MH or DMH. I wanted to share my experiences with my fellow men, but felt that before doing so I had to have more evidence, so I decided to set up a research project to study these effects. Dr. Berndt Sjöberg of the Astra Works in Sjödertaelje, Sweden, kindly supplied me with the necessary amounts of MH which he had synthesized for me. But I soon found myself unable to initiate this study which demanded not only MH, but also money (which I do not have) and a very great deal of time (which I cannot afford). Such studies demand special organization. Resistance to colds is only one of the numerous problems of great import to human health that could be solved only on a statistical scale by a specific organization, but unfortunately no such organization exists.

I would like to close this chapter with a quaint story. While I was isolating ascorbic acid in Hungary, a patient was admitted to the medical clinic of my University with extensive subcutaneous bleeding. Since such bleeding is a classic symptom of scurvy, on advice of Professor St. Rusznyak, my impure preparation of ascorbic acid was injected into the patient, whereupon the bleeding stopped. After I crystallized ascorbic acid another patient with the same complaint was treated with the pure vitamin C. It had no effect. I had a hunch that the action in the first patient may have been due to the flavones present as impurity and so several similar cases were treated later with flavones with excellent results. It seemed possible that the flavones too were vitamins. I was not sure of this so, tentatively, I called them "vitamin P," using the letter P because it was toward the end of the ABC's, still far from the letters used in vitaminology. In the event that I was wrong, the name could be dropped without causing any difficulty. I also chose "P" because the name of most pleasant things, in Hungarian, begin with the letter P.

The idea that flavones are vitamins was rejected in the United States, no deficiency disease having been found which could have been ascribed to their lack. Nevertheless, extensive experimental research work was conducted in Hungary which definitely showed flavones to be vitamins. The discrepancy is easy to explain. Orange juice, widely used in the United States, is rich in flavones and so are most vegetables. So, for Americans, flavones are not needed as vitamins; there is no need to supplement the diet with them. In Hungary, after the war, citrus fruits were a rarity, vegetables were in poor supply, so the lack of flavones caused problems. Flavones are vitamins, their deficiency leading to cerebral edema and bleeding which can be cured with their administration.

REFERENCES

Bungeberge de Jongh, H. L., Klaar, W. J., and Vilegenhart, J. A. (1953). *Nature (London)* 172, 402.

Cosgrove, D. J., Daniels, D. G. H., Greer, E. N., Hutchinson, J. B., Moran, T., and Whitehead, J. K. (1952a). *Nature (London)* 169, 966.

Cosgrove, D. J., Daniels, D. G. H., Whitehead, J. K., and Goulde, J. D. S. (1952b). *J. Chem. Soc., London* p. 4821.

Goodall, F. (1965). *Arch. Biochem. Biophys.* 112, 403.

Goodall, F. (1966). *Science* 152, 356.

Haworth, W. N., and Szent-Györgyi, A. (1933). *Nature (London)* 131, 24.

Lundegaerdh, H. (1939). *Planta* 29, 419.

Pauling, L. (1970). "Vitamin C and the Common Cold." Freeman, San Francisco, California.

Szent-Györgyi, A. (1925). *Biochem. Z.* 162, 399.

Szent-Györgyi, A. (1928). *Biochem. J.* 22, 1387.

Szent-Györgyi, A. (1930). *Science* 72, 125.

Szent-Györgyi, A. (1931). *J. Biol. Chem.* 90, 385.

Szent-Györgyi, A. (1970). "Electrons, Defense and Regulation. W. O. Atwater Memorial Lecture 1970." Agr. Res. Serv. U.S. Dept. of Agriculture, Washington, D.C.

Vouataz, L. (1950). *Helv. Chim. Acta* 33, 443.

VII

ANIMAL DEFENSE MECHANISMS

1. GENERAL REMARKS

Animals, having a blood circulation, can mobilize against microbial invasion the immunological defense mechanism of their whole bodies and need not depend on local reactions. It is only mechanical damage that they have to correct locally by cell division to heal a wound or regenerate a severed part. The classic example of this defense mechanism is the rat's liver. Eight days after two-thirds of the liver has been removed, it is completely regenerated. Removal of a portion of the liver initiated an explosive outburst of cell division. This enormous growth potential could not have been induced just by the surgery but must have been inherent in the liver cells as an attribute of life.

The first simple anaerobic organisms could proliferate freely, limited only by environmental factors such as the availability of food. When a richer energy source was found in light and oxidation, the cells differentiated and more complex multicellular structures resulted; cell division had to be suppressed in the interest of harmony of the organism as a whole. Nevertheless, the organism retained its potential for growth, releasing it only when cell division was needed. Highly differentiated cells, such as nerve cells, have lost this potential.

These facts are not only fascinating but are also of great importance to medicine. When the cell loses its ability to arrest cell division, cancer results.

On the following pages I will be concerned with growth regulation as a defense mechanism which corrects mechanical injury. We can assume that this mechanism follows the general pattern of defense mechanisms: the damage elicits its own correction. The disorder, created by the damage, activates an enzyme system that triggers the correction, which, in the present case, is cell division.

Analyzing these reactions, we can follow an empirical route and ask whether cells contain inhibitors of growth which can keep the cell at rest. In this case cell division could be induced by the enzymic destruction of the inhibitors. We can also follow a theoretical path and inquire into parameters which can arrest proliferation. In this case cell division would have to be elicited by a change in parameters.

The chemical machinery of biological oxidation is a rather bulky one. It involves solid state and structure. When the cell divides it has to rearrange its whole interior, which demands mobility. To achieve this the cell has to dispense with its bulky oxidative machinery, dismount it, and revert to the simple and older anaerobic energy production. The faster the cell divides the more completely it will have to revert to fermentation from oxidation as shown by O. Warburg (1966). Cell division and anaerobic fermentation are coupled processes.

2. RETINE

Many years ago Jane McLaughlin and I found that extracts of the thymus gland inhibited cell division in mice. We used inoculated cancer cells, which grow fast, as test material, and in order to observe inhibition of cell division one needs rapid growth. We called the hypothetic inhibitory substance "retine." Later, in our futile attempts at isolating this substance, we were joined by Hegyeli (Szent-Györgyi et al., 1962–1963a,b; Hegyel et al., 1963; see also Szent-Györgyi, 1965; Szent-Györgyi et al., 1967; Együd et al., 1967). Similar inhibitory actions were also observed by Parshley (1965) with various tissue extracts, but no attempts were made at isolating them.

The higher the growth potential of a cell the more inhibitor it needs. Impressed by its enormous regenerative power, we have recently concentrated on the liver and found that the extracts of 1 gram of a mouse's liver, injected in two doses per day, could greatly reduce the growth of induced cancer in 25-gram mice, and concluded that if the extract of 1 gram of liver could reduce growth in the whole animal, it could certainly reduce it in the liver itself, keeping its cells in interphase.

We were not the first to find such inhibitory activity with liver extracts. In 1956, Glinos found rat serum inhibitory for growth in tissue culture. The serum of rats with partial liver extirpation did not possess this inhibitory action. Two years later, Stich and Florian (1958) found that homogenates of liver inhibited mitosis in regenerating liver *in vivo*. Suspensions of regenerating liver had only slight inhibitory action. They also found normal serum inhibitory but not that of animals with regenerating liver. Homogenates of other organs did not exhibit this activity, so they concluded that an organ-specific inhibitor existed. In serum this inhibitor was destroyed in 24 hours. It prevented the initiation of mitosis, but was unable to arrest it once it started.

In 1960, Herbut and Kraemer found the growth of implanted lymphosarcoma in mice strongly inhibited by liver extracts. They obtained complete inhibition with saline extracts of guinea pig liver; marked, but incomplete, inhibition with extracts of sheep, hog, and rabbit liver; slight and inconsistent inhibition with horse and bovine liver. Earlier, they had found inhibitory activity using extracts of guinea pig blood, and located the active substance in the globulin fraction. The substance could be extracted with methanol and withstood heating to 66°C for 30 minutes.

In 1961, Nakahara and Fukuoka found that ascites cells, after having been suspended in aqueous liver extract for ½–1 hour, no longer produced cancer, and speculated about the possible importance of a "carcinostatic liver factor" (1963; Nakabara *et al.*, 1963). They found that the active material was not destroyed by heat, not even by autoclaving, and was specific for the liver. Extracts of other organs were inactive. Later (Nakahara and Fukuoka, 1965; Nakahara *et al.*, 1965) they injected their active extracts into tumor-bearing mice and found that the extracts had no effect on the rate of growth and so discarded their earlier ideas about the importance of the extract on the cancer-host relation.

The experiments of Glinos and Stich and Florian suggested that in dividing cells the inhibitor was absent and so its destruction might have been responsible for the onset of cell division. Nakahara and Fukuoka found that liver extracts of tumor-bearing animals were inactive as were the extracts of hepatomas.

Mouse or rat liver, owing to its small size, holds little promise of isolation and identification of a substance possibly present in low concentration. This made us, Jane A. McLaughlin and myself, shift to calf's liver. The inhibitory action of our extracts* was judged by the inhibitory action on the growth of implanted tumor in Swiss albino mice. We injected about 100 or 5 $\times$ 10^6 Sarcoma 180 or Krebs 2 ascites cells into 25-gram mice on one side behind the scapula.† The tumors were excised 8 days later and weighed. Injections of liver extracts started one day after inoculation. Extracts were injected twice daily intraperitoneally in a volume of 0.25 ml. The results of a representative experiment with Krebs 2 carcinoma are reproduced in Fig. 22. Sarcoma 180 gave similar results. The points indicate the weights of the individual tumors. The first column shows the weights of the tumors of untreated animals

---

*Extracts were prepared by suspending 1 gram of minced liver in 2.5 ml of water; then 0.0035 ml acetic acid were added and an equal volume of methanol, the precipitate eliminated, the fluid concentrated *in vacuo* and treated again with higher concentrations of methanol. Eventually the methanol was eliminated from the clear supernate by distillation.

†In earlier experiments we injected 5 million cells. With the greater number of inoculated cells the inhibiting actions obtained may be weaker, and the scattering stronger. In using 100 cells we followed the procedure used in G. Klein's laboratory (Karolinska Institute).

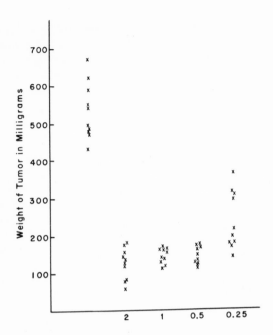

Fig. 22. Inhibitory effect of liver extract on the growth of inoculated Krebs 2 carcinoma. The points indicate the weights of the individual tumors; the numbers the grams of liver extract ingested daily, divided into two doses. The first column represents the weights of the tumors of untreated animals receiving saline injections only; the remaining columns represent those of animals treated with liver extracts.

receiving saline injections only, the next columns those of animals treated with extracts of 2, 1, 1/2, or 1/4 gram of the liver. As is shown in the figure, even the extract of 1/2 gram of liver contains enough inhibitor to suppress tumor growth when administered daily. The percents of inhibition were 77, 74, and 73.

These results had an unusual feature. As is shown in Fig.22, the extract of 2 grams of liver, administered daily, had the same effect as the extract of 1 or 1/2 gram of liver. The most probable explanation is that the inhibitor reacted with a tissue constituent in stoichiometric relations. Once this constituent was saturated an excess had no additional action.

The inhibitory substance was not precipitated by methanol, was partially precipitated by ethanol, while most of it was precipitated by acetone. The partially purified extracts could be stored at 0°C for longer periods without loss of activity. They withstood moderate heating. The inhibitor was retained by membranes cutting off at 200 grams molecular weight (Dow Hollow Fiber Dialyzer HFO-1), was partially retained by membranes cutting off at 500 grams molecular weight (Amicon Diaflo UM-05), and passed through membranes cutting off at 1000 grams molecular weight (Amicon Diaflo UM-2). This suggests a substance of low molecular weight, about 500 grams.

Our extracts also displayed bacteriostatic action, arresting the growth of bacteria; we never lost an extract due to putrefaction. There was no loss of weight. The extracts had no general toxicity.

The isolation of the active substance appears to be difficult, due partly to the properties of the substance that, hitherto, had revealed no specific reactivity. It readily absorbs to various precipitates. The difficulty is also due to the cumbersome, innacurate, and time-consuming nature of the measurement of tumor growth. Attempts at isolation are being continued.

3. KETONE ALDEHYDES

Proliferation of microorganisms is suppressed in the vegetable kingdom by hydrogen or electron acceptors which deprive the invading organisms of their active H's or electrons and bind those of the medium, or interfere with the catalytic function of SH groups either by oxidizing their H's or by forming compounds with them.

As has been shown before, the most common acceptor is the $C=O$. Owing to their excessive activity, aromatic $C=O$'s, that is quinones, cannot be used by animal tissues as acceptors. A $C=O$ group at the end of an acyclic molecule leads to the formation of an aldehyde, while inside the chain the $C=O$ leads to a ketone. Ketones are thermodynamically favorable, but are not very reactive. Aldehydes are more reactive but, owing to their hydration, are thermodynamically less favorable. They can act not only as acceptors but can also form hemimercaptales with the catalytic SH groups. The inhibitory action of aldehydes on malignant growth was noted some thirty years ago by Strong and Whitney (1938). Nevertheless, the aldehydes never occupied a place on the list of useful oncostatic agents. The action of lower aldehydes is not specific, while the higher aldehydes are less reactive. The activity of an aldehyde can be increased by introducing into the molecule a second, conjugated double bond in the $a-\beta$ position. The $C=C$ bond is unfit for this purpose because the β C atom is prone to addition, which makes it poisonous and unfit for a reversible, catalytic function. A second double bond can be introduced also by an additional $C=O$ group. A second aldehydic $C=O$ gives us glyoxal (Fig. 23). The introduction of a ketonic $C=O$ into an aldehyde in an a position leads to a ketone aldehyde, a "glyoxal derivative," the simplest being methylglyoxal (Fig. 24). Underwood (1956) was the first to note the antiviral action of an aldoketone (β-isopropyl-a-keto-butyraldehyde). Subsequently, Tiffany $et\ al.$ (1957) synthesized a number of different ketone aldehydes and

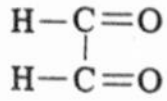

$$H-C=O$$
$$H-C=O$$

Fig. 23. Glyoxal.

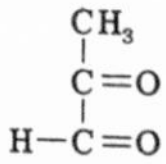

$$CH_3$$
$$C=O$$
$$H-C=O$$

Fig. 24. Methylglyoxal.

found them active. One of them, kethoxal (β-ethoxy-α-ketobutyraldehyde) is on the market, and is a well-known oncostatic agent.

What makes the glyoxal derivatives exciting for the biologist is the fact that, as far as we know, all cells contain a very active enzyme, glyoxalase, for the inactivation of ketoaldehydes. It turns them into the corresponding hydroxy acids, e.g., methylglyoxal into lactic acid (Fig. 25).

Glyoxalase was discovered by Neuberg and by Dakin and Dudley in 1913. It is one of the most active enzymes. As stated by Dakin and Dudley, a liver extract can convert twice its own weight of methylglyoxal in 15 minutes. Since Nature does not indulge in luxuries, if there is such a highly active and widely spread enzyme for the inactivation of ketoaldehydes then it must have something important to do. In the first half of the century many of the leading biochemists worked on this enzyme but since nobody could find a substrate for it interest in it gradually died.

Glyoxalase is not an enzyme but an enzymic system which involves two enzymes, glyoxalase I and II, and a coenzyme, SH-glutathione (Lohman, 1932). The function of this complex is represented in Fig.

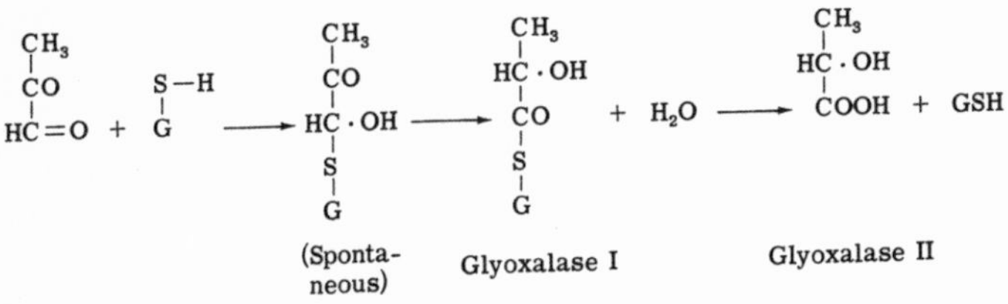

$$CH_3$$
$$|$$
$$CO$$
$$|$$
$$HC{=}O \ + \ \overset{S-H}{\underset{G}{|}} \ \longrightarrow \ \overset{CH_3}{\underset{G}{\underset{|}{\overset{|}{CO}}}} \ \overset{|}{HC \cdot OH} \ \longrightarrow \ \overset{CH_3}{\underset{G}{\underset{|}{\overset{|}{CO}}}} \ \overset{|}{HC \cdot OH} \ + \ H_2O \ \longrightarrow \ \overset{CH_3}{\underset{COOH}{\overset{|}{HC \cdot OH}}} \ + \ GSH$$

(Sponta-   Glyoxalase I   Glyoxalase II
neous)

Fig. 25. Reaction mechanism of glyoxalase.

25. In the first step the SH forms, spontaneously, a hemimercaptal with the aldehyde (Kühnau, 1931; Jowett and Quastel, 1932), then glyoxalase I shifts two H atoms to the β C atom, whereupon glyoxalase II splits the complex, recovering the SH-glutathione which acts as coenzyme. This mechanism was cleared up by Racker (1952) and Crook and Law (1952).

The function of the glyoxalase, evidently, is to inactivate a ketone aldehyde, but the substrate of this enzyme is not the ketone aldehyde itself, but the hemimercaptal it forms with glutathione, as emphasized by Jowett and Quastel. Glyoxalase I is much more active than glyoxalase II.* It belongs to the most active metabolic enzymes.

The ubiquitous glyoxalase explains why no free glyoxal derivatives could be found in tissues: they have to be decomposed. Nevertheless, the existence of glyoxalase suggested an important biological role for ketoaldehydes. I started to study them with L. G. Együd who left the Lister Institute of London to join me. He found methylglyoxal in one of our "retine" preparations (1965). Eventually, we were led to the assumption that

*The correct name for glyoxalase I is S-lactoylglutathione-methylglyoxal-lyase (4.4.1.5); for II it is S-2-hydroxyacyl-glutathione-hydrolase (3.1.2.6).

an aldoketone may be involved in the regulation of cell division (Együd and Szent-Györgyi, 1966a,b).

Glyoxalase is not the only enzyme which can decompose and inactivate glyoxal derivatives. C. Monder discovered a "ketone aldehyde dehydrase" (KAD) which could oxidize methylglyoxal to pyruvic acid (1965, 1967). Methylglyoxal can easily be formed, and can be inactivated both by glyoxalase and the KAD, which makes it easy to believe that with its formation and decomposition it regulates cell division. Ketone aldehydes and pyruvic acid are H or electron acceptors, while lactic acid is, in the presence of latic dehydrase, a good H donor. Ketone aldehydes may thus play an important role in shifting the D/A quotient and the balance between fermentation and oxidation.

Monder also found in tissue an inhibitor for KAD. Should ketone aldehydes be responsible for keeping the cell in the interphase, then an enzyme such as KAD, which can decompose or inactivate them, should be capable of initiating cell division, while an inhibitor, capable of inactivating this enzyme, should inhibit cell division, act as "retine." So the glyoxalase and KAD and its inhibitor may together form a system capable of a subtle regulation of cell division.

That my liver extracts contained no free ketone aldehydes could be shown by the addition of dinitrophenylhydrazine which should immediately have formed a deep red precipitate with the ketone aldehydes present. No such precipitate formed. However, within a couple of days such a precipitate did form which contained five main fractions, the main fraction being the dinitrophenylhydrazone of methylglyoxal. This suggests that our extracts contained ketoaldehyde complexes which were split by the strong acidic reaction of the reagent. Kato (1960) isolated from liver dinitrophenylhydrazones of eighteen different carbonyls, out of which, with his mastery of the field, he identified five.

Kenny and Sparkes (1968, 1969) isolated from their tissue cultures a bacteriostatic ketoaldehyde which they identified as a-keto-γ-hydroxybutyric aldehyde.

4. INHIBITION OF CELL DIVISION BY KETONE ALDEHYDES

If ketone aldehydes are regulators they should be able to inhibit cell division reversibly without damaging the cell. In order to decide whether they are capable of doing so, the homolog series of methylglyoxal was synthesized up to C_{13} (Együd and Szent-Györgyi, 1966 a). Also, various cyclic derivatives were synthesized. The substances were tested for their action on the growth of Escherichia coli. With the higher aliphatic members, over C_6, there was some difficulty owing to poor solubility, but nevertheless the results showed that except for the C_{12}, methylglyoxal homologs can suppress cell division completely in millimolar concentration, and do this without damaging the cells since the bacteria readily resumed growth as soon as the aldoketone was decomposed by the glyoxalase present, or was inactivated, be it by SH compounds or ethylenediamine. SH compounds were especially active when they contained two SH's in neighboring positions which enabled them to react simultaneously with both CO groups of the ketone aldehyde.* This protecting action

*Since on oxidation the "fixed SH's" (the SH linked to protein) form disulfides (S–S), it follows that the reacting SH groups must be side by side, being located on the same peptide chain or on different chains held in close proximity. This explains why the substances containing two SH groups in neighboring position give the strongest protection against high-energy radiation. This may also explain the high biological activity of ketoaldehydes which also have the two CO's in neighboring position.

of dithiols paralleled their protecting action against radiation damage. The action of the ketone aldehydes depends less on their absolute concentration than on the relation between their quantity and the number of cells acted upon, suggesting a stoichiometric interaction. These observations were later extended to other systems such as fertilized sea urchin and frogs' eggs, flagellates, germinating seeds, normal or cancerous mammalian cells, viruses, and ascites tumor. The results were similar.

Protein synthesis was found strongly suppressed by the glyoxal derivatives, as measured by the incorporation of the radioactively labeled lysine-^{14}C (Együd and Szent-Györgyi, 1966b). The incorporation of labeled thymidine-^{14}C and uracil-^{14}C were only moderately inhibited, indicating that the inhibition of DNA and RNA synthesis was not the limiting factor. Since proteins are needed for the synthesis of nucleic acid, it is evident that protein synthesis cannot be inhibited without inhibiting, to some extent, the synthesis of nucleic acid. During the inhibition of protein synthesis the nucleic acids accumulated, and after cessation of the inhibition the incorporation of lysine proceeded for some time at an increased rate. All this suggested that aldoketones arrested cell division by interacting with SH groups. By doing so they shut off the H pool and arrest protein synthesis on the ribosome level (Otsuka and Együd, 1968b).

Shun-ichi Hata's (1970) measurements indicated that the specific action of ketone aldehydes on growth could not be explained solely by their electron-acceptor ability. He measured the polarographic potential of a series of acceptors, and calculated their ionization potential and electron affinity. When plotting the electron affinities against growth inhibition, he found that most of

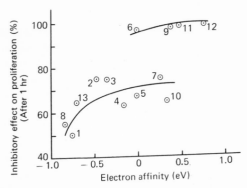

Fig. 26. Relation between the inhibitory effect on proliferation of *E. Coli* and electron affinity. 1, Propionaldehyde; 2, *n*-butylaldehyde; 3, valeraldehyde; 4, heptaldehyde; 5, crotonaldehyde; 6, methylvinylketone; 7, mesityloxide; 8, glyoxal; 9, methylglyoxal; 10, acetophenone; 11, ethylglyoxal; 12, propylglyoxal; 13, 2-butanone.

the substances studied fell on a line and formed a coherent group. The ketone aldehydes did not fit in and formed a separate group in which the growth inhibition was stronger than corresponded to the electron affinity (Fig. 26). This difference may have been due to its ability to form hemimercaptals with SH.

It had been known for some time that ketone aldehydes also had antiviral activity, an effect due to direct action on the virus. Staehelin (1959) found that glyoxal and a glyoxal derivative, "kethoxal" (β-ethoxy-a-ketobutyraldehyde) interacted with the guanine of nucleic acid. Litt and Hancock (1967) showed that kethoxal reacted preferentially with single-stranded RNA.* Shapiro and Hachman (1966) isolated and analyzed the reaction product of guanine and methylglyoxal.

*These observations make it seem worthwhile to try to sterilize blood to be used for transfusion with methylglyoxal. If 0.01

95

It is easy to demonstrate an interaction of methylglyoxal and guanine. Guanine–HCl is very insoluble. It
dissolves in boiling water but readily precipitates on
cooling. If two equivalents of methylglyoxal are present,
it remains in solution and the solution can be completely dried to a glass, behavior specific for guanine.

All this makes it seem possible that an interaction of
a glyoxal derivative with single-stranded RNA may also
play a role in the inhibition, though the main factor is
due to the blocking of SH groups.

Evidence that glyoxal derivatives do interact *in vivo*
with the SH groups has been found by Ashwood-Smith
and his associates (1967). As is generally known, SH
protects against high-energy radiation. Should glyoxal
derivatives interact with SH in tissues, they should increase sensitivity toward X-rays. That they actually do
so has been shown by the British authors.

Glutathione, though having an SH group, is unable to
release the inhibition of cell division induced by ketone
aldehydes. The ketone aldehydes thus have a greater
affinity for the fixed SH of tissues than for the SH of
glutathione, and so the ketone aldehydes will be able to
inhibit cell division in the presence of glutathione also.

5. KETONE ALDEHYDE COMPLEXES

Though ketone aldehydes inhibit cell division, they
cannot be expected to exert oncostatic action if injected

M methylglyoxal plus 0.01 M NaHCO$_3$ were added to the blood,
the glyoxal derivative could be expected to inactivate the hepa-
titis virus present. Subsequently the glyoxalase present would
transform the methylglyoxal into the harmless lactic acid, which
would be neutralized by the NaHCO$_3$.

Underwood and Weed (1956) had proposed earlier the use of
glyoxal and related compounds for blood sterilization.

into an animal some distance from the tumor since they are readily decomposed by the glyoxalase of the blood. To exert such an action they would have to be injected into the tumor. Együd and Szent-Györgyi (1968) injected methylglyoxal into the ascites tumors of mice and found that a considerable number of the animals were cured. Apple and Greenberg (1967) did likewise and obtained similar results. Their results were the first to be published.

Ketone aldehydes could be expected to exert a general oncostatic action *in vivo* if protected from the glyoxalase. Such protection could be given by complexing the aldehydic CO in a reversible way with some group which could be split off by the tumor tissue, but not by the blood. Led by these assumptions, Együd synthesized a series of ketone aldehyde complexes in which the aldehydic CO formed a complex with various amines. He obtained considerable oncostatic action. I found that methylglyoxal formed with hydrazine various hydrazones. In the presence of an excess of methylglyoxal it formed a bismethylglyoxalhydrazone (Fig. 27) which exhibited considerable oncostatic action (Fig. 28). The treatment, in this experiment, started on the day following the injection of 5 million Sarcoma 180 ascites cells.

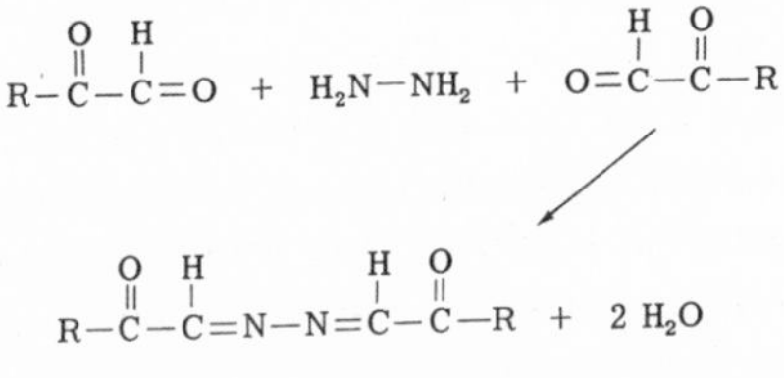

$$R-\overset{O}{\overset{\|}{C}}-\overset{H}{\overset{|}{C}}=O \; + \; H_2N-NH_2 \; + \; O=\overset{H}{\overset{|}{C}}-\overset{O}{\overset{\|}{C}}-R$$

$$R-\overset{O}{\overset{\|}{C}}-\overset{H}{\overset{|}{C}}=N-N=\overset{H}{\overset{|}{C}}-\overset{O}{\overset{\|}{C}}-R \; + \; 2\,H_2O$$

Fig. 27. Interaction of two molecules of a ketone aldehyde with one molecule of hydrazine.

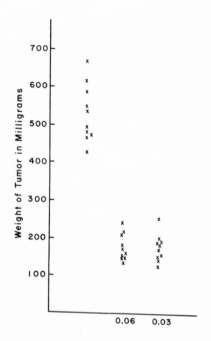

Fig. 28. Inhibitory action of bismethylglyoxalhydrazone on the growth of Sarcoma 180. Left column, placebo; middle column, 0.06 M hydrazone; right column, 0.03 M hydrazone; 0.25 ml injected intraperitoneally twice daily.

We were not the first to experiment in this field. That hydrazones of ketone aldehydes have a carcinostatic action was well known. The pioneers in this field, for which there is an extensive literature, were French and Freelander (1958; Freelander and French, 1958) who studied bissemicarbazides, bisthiosemicarbazides, and bisguanidylhydrazones of ketone aldehydes. They attributed the carcinostatic activity of these compounds to their ability to form chelates with metals and not to the specific action of the aldoketone.

That hydrazone derivatives occupy no more important place in our oncostatic armory is probably due to the poisonous nature of hydrazine or its derivatives. Freelander and French found their compounds more active on per os administration than on intraperitoneal injection. This cannot be said about the tested bismethylglyoxalhydrazine which was found less active if administered per os, possibly being split by HCl in the stomach.

6. A THEORY OF REGULATION

The evidence presented suggests that the cells of multicellular organisms are kept at rest by a specific substance which is called "retine," and, possibly, is a ketone aldehyde derivative. If so, then cell division has to depend on the balance of the production and destruction of this substance. It may be this balance which is disturbed when cell division goes awry due to damage and its resulting disorder. What role glyoxalase plays is not yet known. Perhaps it functions only as a scavenger, eliminating free ketone aldehydes.

The experimentation suggested by this theory is fourfold: How is retine produced? What is the mechanism of its action? How is it inactivated? And, finally: What is the relation of this system to cellular metabolism?

As suggested by the classic studies of O. Warburg (1966), the cell has two extreme states: the proliferative anaerobic state and the resting aerobic state. When life originated there was no O_2 in the atmosphere, and so life must have started with the proliferative anaerobic state. Later, with the appearance of oxygen and the multicellular state, proliferation had to be suppressed. Perhaps this is achieved by products of the oxidative metabolism, H+ acceptors, which bound the H, the energy source of anaerobic proliferation, and inactivated the SH, the tap on the H pool. The dividing cell, in order to achieve a greater mobility, has to dismount the bulky oxidative mechanism and shunt back, partly or wholly, to the anaerobic energy production. The anaerobic state and proliferation are coupled. The cell can also be forced into this anaerobic state by withholding O_2. If this anaerobic state lasts too long or is induced repeatedly, the proliferative anaerobic state may become constitutive, and the cell may be unable to shift back to the aerobic resisting state even if O_2 is reintroduced. In this case proliferation may be arrested by the introduction of H acceptors, which cannot be produced by an anaerobic metabolism.

In carcinogenesis we have to distinguish between two steps. The first is physiological. It is the shift from the aerobic resting state to the anaerobic proliferative state. The second, pathological step is the stabilization of this state, its becoming constitutive. This step may be an expression of the general tendency of unused systems to deteriorate as is discussed in Chapter IV.

7. FAILURES AND RED HERRINGS*

I ran into my first "red herring" when I began to speculate about the relationship of fluorescence and growth inhibition, and was impressed by the presence of a very small quantity of a strongly fluorescent matter in my tissue extracts. The quantity was very small, but the brilliance of the blue fluorescence was striking. So I isolated the substance. It was a "tour de force." I sent the crystals to my friends at Merck & Company request-

---

*The reading of this section is not required for understanding or continuity.

ing that they analyze them for me. I soon obtained the answer, which revealed that I had not extracted this substance from the tissues but from my rubber tubing to which it had been added by the manufacturer as a stabilizer. It could be bought for a dollar per pound. It was N-phenyl-β-naphthylamine.

It was much later that I was led to think about SH and charge transfer. If cell division depends on the presence of small quantities of SH, one may expect that cell division could be inhibited by keeping this SH in an oxidized condition by the permanent addition of low concentrations of a weak electron acceptor. I approached this problem in different ways. I injected mice with ascorbic acid plus ascorbic acid oxidase, hoping to maintain a low concentration of dehydroascorbic acid which is a weak acceptor or oxidizing agent. In another set of experiments, I injected various polyphenols plus a polyphenol oxidase, hoping that the low concentration of quinone produced might depress cell division. In a third set of experiments, I injected methoxyhydroquinone and manganese, hoping that the low concentration of quinone produced by the autoxidation of the phenol would achieve the desired effect. Inhibitions were obtained, but the simultaneous loss of weight of the animals indicated that the effect was not specific, but was due to general toxicity.

In order to maintain an electron acceptor without enzyme, I searched for a highly autoxidizable polyphenol. This I found in mushrooms, *Agaricus bifidus*, the cultivated form of *Agaricus campestris*. This species contains an excessively autoxidizable polyphenol in very small quantity. I used thousands of pounds of mushrooms to isolate this substance in crystals. It was a fascinating substance. In a slightly alkaline solution, it

was rapidly oxidized to a red-brown quinone or meri-
quinone. Autoxidation was increased by manganese.
With ferric ions it produced a wonderful blue color at an
alkaline pH. My friend, Dr. M. Tishler of Merck &
Company, to whom I am most grateful for support
throughout my years in America, came to my aid, as did
his associates. Their analysis showed my substance to be
a 4-(L-γ-glutamylamino)pyrocatechol. However fascinating, it did not cure cancer.

After all these disappointments, I pinned my hope on
the isolation of retine, which I supposed to be a ketone
aldehyde. The possibility occurred to me that I was
unable to find this substance because it formed a compound with the SH groups present. So I treated liver
extracts with arsenious acid hoping to inactivate all SH
as an As compound. To my delight I found a great
quantity of a ketone aldehyde which I isolated in crystals as 2,4-dinitrophenylosazone. My friend Professor G.
Fodor, with his associate J. P. Sachetto, found it to be a
carbohydrate derivative, a glucosulose, the same substance Kato (1960) found in soybeans earlier. This,
finally, appeared to be the solution to the problem. Professor Fodor and Sachetto synthesized the free substance
which I tested biologically (Fodor et al., 1967). It was
inactive and was not attacked by glyoxalase.

Baker and Együd (1968) found that this ketone aldehyde was produced postmortem. Otsuka and Együd
(1968a) found that glucose and various amino acids, in
the presence of inorganic phosphate, yielded a compound which, in the presence of As_2O_3 decomposed to
deoxyglucosulose. A decade earlier, Borsook and his
associates (1955) discovered that liver contained a re-
markably large quantity of fructose amino acid; Heyns
and Paulsen (1959) found that it was formed after
death, on storage. Professor Fodor suggested that it may

have been the precursor of our ketone aldehyde. He and Sachetto synthesized L-fructosylvaline and glycine, which they treated with As_2O_3 in the same way I had treated my extracts. It yielded 3-deoxyhexosulose, our ketone aldehyde. The series of reactions by which this substance was formed was rather unexpected. If glucose unites with an amino acid, a so-called "Amadori rearrangement" takes place, and instead of a glucose amino acid one obtains a fructose amino acid. This compound is split by arsenic, and the splitting involves a second rearrangement, a shift of oxygen from C-3 to C-1. The ketone function of our deoxyglucosulose was thus provided by the Amadori rearrangement, while the aldehyde function was produced by the action of arsenic. All these reactions were very interesting and unexpected, but left me with the sad conclusion that I had run into another "red herring."

The last "red herring" was encountered when we embarked on the polarographic study of our oncostatic extracts with Dr. Shun-ichi Hata (Hata et al., 1971). We found two reductive peaks, one around 1.4 and one around 1.7 eV. It turned out that these peaks were not related to the inhibitory action of our extracts. This last herring was not completely red since it led to the discovery of a third peak at 1.9 eV which was given by a substance specific for cancer and for regenerating of rapidly growing tissue. It is hoped that its further study will contribute to the understanding of the chemical mechanism of rapid growth and cell division.

REFERENCES

Apple, A. M., and Greenberg, M. (1967). *Cancer Chemother. Rep.* **51**, 455.

Ashwood-Smith, M. J., Robinson, D. M., Barnes, J. H., and Bridges, B. A. (1967). *Nature (London)* **216**, 137.

Baker, N., and Együd, L. G. (1968). Biochim. Biophys. Acta 165, 293.

Borsook, H., Abrams, A., and Lowy, P. H. (1955). J. Biol. Chem. 215, 111.

Crook, E. M., and Law, K. (1952). Biochem. J. 52, 492.

Dakin, H. D., and Dudley, H. W. (1913). J. Biol. Chem. 14, 155 and 423.

Együd, L. G. (1965). Proc. Nat. Acad. Sci. U.S. 54, 200.

Együd, L. G., and Szent-Györgyi, A. (1966a). Proc. Nat. Acad. Sci. U.S. 55, 388.

Együd, L. G., and Szent-Györgyi, A. (1966b). Proc. Nat. Acad. Sci. U.S. 56, 203.

Együd, L. G., and Szent-Györgyi, A. (1968). Science 160, 1140.

Együd, L. G., McLaughlin, J. A., and Szent-Györgyi, A. (1967). Proc. Nat. Acad. Sci. U.S. 57, 1422.

Fodor, G., Sachetto, J. P., and Szent-Györgyi, A. (1967). Proc. Nat. Acad. Sci. U.S. 57, 1644.

Freelander, B. L., and French, F. A. (1958). Cancer Res. 18, 1286.

French, F. A., and Freelander, B. L. (1958). Cancer Res. 18, 172 and 1290.

Glinos, A. D. (1956). Science 123, 673.

Glinos, A. D., and Grey, G. (1957). Proc. Soc. Exp. Biol. Med. 80, 421.

Hata, S.-i. (1970). Bioenergetics 1, 325.

Hata, S.-i., Együd, L. G., and Szent-Györgyi, A. (1971). Proc. Nat. Acad. Sci. U.S. 68, 2992.

Hegyeli, A., McLaughlin, J. A., and Szent-Györgyi, A. (1963). Proc. Nat. Acad. Sci. U.S. 49, 220.

Herbut, P. E., and Kraemer, W. H. (1960). Amer. J. Pathol. 36, 105.

Herbut, P. E., Tsaltas, T. T., and Kraemer, W. P. (1963). Amer. J. Clin. Pathol. 39, 298.

Heyns, K., and Paulsen, H. (1959). Justus Liebigs Ann. Chem. 622, 160.

Jowett, M., and Quastel, J. H. (1932). Biochem. J. 27, 486.

Kato, H. (1960). Agr. Biol. Chem. 24, 1.

Kato, H., Tsusaka, N., and Fujimaki, M. (1970). Agr. Biol. Chem. 33, 2541.

Kenny, C. P., and Sparkes, B. G. (1968). Science 161, 1344.

Kenny, C. P., and Sparkes, B. G. (1969). Proc. Nat. Acad. Sci. U.S. 64, 920.

Kühnau, J. (1931). *Biochem. Z.* **243**, 14.

Litt, M., and Hancock, V. (1967). *Biochemistry* **6**, 1848.

Lohman, K. (1932). *Biochem. Z.* **254**, 332.

Monder, C. (1965). *Biochim. Biophys. Acta* **99**, 573.

Monder, C. (1967). *J. Biol. Chem.* **242**, 4603.

Nakahara, W., and Fukuoka, F. (1961). *Gann* **52**, 197.

Nakahara, W., and Fukuoka, F. (1963). *Naturwissenschaften* **50**, 406.

Nakahara, W., and Fukuoka, F. (1965). *Gann* **56**, 87.

Nakahara, W., Fukuoka, F., Sugimura, T., and Hozumi, M. (1963). *Naturwissenschaften* **50**, 406.

Nakahara, W., Fukuoka, F., Maeda, Y., Tokusen, R., and Tsuda, M. *Gann* **56**, 87.

Neuberg, C. (1913). *Biochem. Z.* **49**, 502.

Otsuka, H., and Együd, L. G. (1968a). *Biochim. Biophys. Acta* **165**, 172.

Otsuka, H., and Együd, L. G. (1968b). *Curr. Mod. Biol.* **2**, 106.

Parshley, M. S. (1965). *Cancer Res.* **25**, 387.

Racker, E. (1952). *Biochim. Biophys. Acta* **9**, 577.

Shapiro, R., and Hachman, J. (1966). *Biochemistry* **5**, 2799.

Staehelin, M. (1959). *Biochim. Biophys. Acta* **31**, 448.

Stich, H. F., and Florian, M. L. (1958). *Can. J. Biochem. Physiol.* **36**, 855.

Strong, L. C., and Whitney, L. F. (1938). *Science* **88**, 111.

Szent-Györgyi, A. (1965). *Science* **149**, 34.

Szent-Györgyi, A. (1967a). *Science* **156**, 543.

Szent-Györgyi, A. (1967b). *Proc. Nat. Acad. Sci. U.S.* **57**, 1642.

Szent-Györgyi, A. (1968). *Perspect. Biol. Med.* **11**, 350.

Szent-Györgyi, A., Hegyeli, A., and McLaughlin, J. A. (1962). *Proc. Nat. Acad. Sci. U.S.* **48**, 1439.

Szent-Györgyi, A., Hegyeli, A., and McLaughlin, J. A. (1963a). *Proc. Nat. Acad. Sci.* **49**, 878.

Szent-Györgyi, A., Hegyeli, A., and McLaughlin, J. A. (1963b). *Science* **140**, 1391.

Szent-Györgyi, A., Együd, L. G., and McLaughlin, J. A. (1967). *Science* **155**, 539.

Tiffany, B. D., Wright, J. B., Moffett, R. B., Heinzelman, R. V., Strube, R. E., Aspergren, B. D., Lincoln, E. H., and White, J. L. (1957). *J. Amer. Chem. Soc.* **79**, 1682.

Underwood, G. E. (1956). *Fifth Nat. Med. Chem. Symp. Amer. Chem. Soc. [Proc.], June, 1956*.

Underwood, G. E., and Weed, S. D. (1956). *Proc. Soc. Exp. Biol. Med.* **93**, 421.

Warburg, O. (1966). "The Prime Cause and Prevention of Cancer," Lecture at the Meeting of Nobel Laureates, June 30, 1966, at Lindau. K. Tiltsch, Würzburg, Germany (English ed. by Dean Burk).

VIII

OBSERVATIONS ON CANCER

1. GENERAL REMARKS

The ability of unlimited proliferation is an attribute of life. In the simplest living forms growth is limited only by environmental factors. In multicellular organisms this potential has to be suppressed in the interest of the organism as a whole. It must be restored when proliferation is needed and again arrested when no longer needed. When the suppression mechanism is out of order and the cell loses its ability to stop proliferation, senseless growth, cancer, results. In this case the problem is not what makes the cell divide, but what has gone wrong with the mechanism so that it cannot stop? A cancer cell is comparable to a car on a slope. If it starts running, the question is not what makes it go, but what's wrong with the brake? As Bullough (1962) puts it "opportunity, not stimulus, is all that is needed for cell division."

If cancer research has not been as successful as it might have been this may be due partly to our "fishing behind the net," asking the wrong question: What makes the cell divide? Another reason may be that we are too eager to cure. Attempts to cure without understanding are shortcuts to failure. Cellular mechanisms are so complex and subtle that trying to find a part which happens to fit into the machine as a brake is a waste of time and money. The only sensible approach is through basic study.

Cancer is a biological, not a clinical or pathological problem. Cancer can be caused by an unlimited number of noxious sources, including viruses. This has tended to make a mystery of cancer and has led to the erroneous idea that cancer is not one disease but many, as many as there are ways to produce it. "There are many ways to Rome," but they may all pass through the same gate. We will understand cancer when we understand normal regulation, and we will understand normal regulation when we can describe it in terms of the basic parameters, energy, entropy, and quantum rules.

2. THE THEORY OF CANCER

As mentioned earlier, the rat's liver can regenerate two-thirds of its total weight within a couple of days. The inflicted cut initiates an explosive outburst of cell division. It would be unphysiological to suppose that the cut has created a new mechanism. A cut is a damage, and a damage can create only disorder, increase entropy.

An increase of entropy can activate dormant systems. It can set free the repressed innate ability of proliferation.

It seems likely that the senseless proliferation, which characterizes cancer, is the expression of a dysfunction, an irreversibly disturbed regulatory mechanism. Oncogenic agents damage the cell and may produce cancer by the disorder they induce.

It has been shown that ketone aldehydes can suppress cell division without damaging the cell. There is no definite evidence yet to prove that ketone aldehydes are involved in the regulation of cell division. However, it has been shown that the actual concentration of ketone aldehydes in tissues may depend on a complex enzymic system and its inhibitors or activators, and so the possibility exists that it is a disturbance of this system which may lead to cancer.*

*New theories can be expected to solve old puzzles. An old puzzle is the difference in the oncogenic activity of a- and β-naphthylamine, a being oncogenic, producing cancer of the bladder, β being harmless. There is no difference in the chemical behavior of the two, the electronic indices of which are similar, except for a slight difference in the dissociation constant (Pullman and Pullman, 1952 and 1963). If aldoketones are key substances in the regulation of cell division, then the two naphthylamines might show a difference in their behavior toward these substances.

If to a 0.05–0.1 M aqueous solution of the hydrochlorides of the two naphthylamines methylglyoxal is added, on storage overnight, a precipitate forms in both, indicating the formation of a complex. The complex of β-naphthylamine is colorless, or has only an unspecific brownish color, while the complex of a-naphthylamine is golden yellow. A brilliant yellow dyestuff is formed, probably a Schiff base formed by the amino group and the aldoketone. The colored complex is soluble in chloroform, but cannot be extracted from its aqueous solution by ether. The naphthylamine, being secreted by the kidney, has to be in contact with the epithelium of the urinary bladder, and may inactivate the ketone aldehyde responsible for keeping the epithelial cells at rest.

3. ON CANCER THERAPY

In our experiments, pursued with J. McLaughlin, the oncostatic action of liver extracts compared favorably with that of two classic oncostatic agents, cytoxan and methotrexate (MTX). This suggests that a therapeutic trial may be indicated even at the present state of the art. A final isolation will create a different, much more favorable situation. Retine is not antigenic; it is a normal cell constituent and is not species specific. If the data obtained hitherto on mice are applied to man, then the extract of 1500 grams of liver, injected daily, should have a therapeutic effect. Possibly, the same effects may be obtained with smaller doses in man than in mice. The activity of drugs, in general, does not depend solely on body weight but also on body surface, or, more precisely, on the relation of body weight to body surface. The larger this quotient the more effective the same drug can be expected to be. In man the quotient is considerably larger than in the mouse.

Another factor that makes the situation less favorable in mice is that no more than two injections can be administered daily and continuous infusion is technically impossible. One injection in the morning and one in the afternoon results in intervals of 8 and 16 hours. Should the retine be excreted or destroyed within 4 hours after injection, this would leave a drug-free interval of 12 hours. In man more continuous application offers no difficulty.

Experiments of Stich and Florian (1958) suggest that retine can prevent cell division but cannot arrest it. Whether retine will be able to cure cancer remains to be seen. Perhaps it will have to be combined with other procedures to achieve a complete cure. A combination with hyperthermy, as proposed by M. V. Ardenne, is one of the possibilities. As has been shown by Ash-wood-Smith and his associates (1967, p. 204), glyoxal derivatives may support high-energy radiation therapy by abolishing the protective action of SH. The reverse may be true too, and high-energy radiation may complete the action of retine. Együd found (unpublished data) that methylglyoxal derivatives injected before irradiation considerably decreased the dose of irradiation necessary to achieve a given inhibition of growth by 50%.

In our experiments, the inhibitory action of a given quantity of retine, administered to tumor-bearing animals, seemed to depend on the size of the inoculum. If 25 million cells were inoculated the inhibition* was 31% (50 animals); with 5 million cells inoculated the inhibition was 41% (60 animals). If the number of the inoculated cells was reduced to 100, the inhibition rose to 63% (160 animals). If only 10 cells were inoculated the inhibition reached 87% (10 animals). These results suggest that if 1 cell only had been inoculated there would have been no growth at all. This is important because cancer or metastasis begins in most cases, probably, with one cell only. Whether this one cell will be able to grow will depend on the actual concentration of retine present. It seems likely that a low retine level favors the onset of growth and cancer, while growth could be completely prevented by keeping the retine level high, which could be achieved with retine administration. This theory found support in the experiment in which we injected into mice the extract of 1 gram of liver twice daily for 5 days. On the fourth day we inoculated the animal, as usual, with 100 cells of Sarcoma 180.

---

*Percent "inhibition" is calculated by $e/c \times 100$, where e is the weight of the tumor of the treated animals and c that of the controls.

Two weeks later we sacrificed the animals. No cancer was found, while in the control animals, receiving saline injections, the tumors grew normally. I am engaged at present in attempts at devising a method for the estimation of retine in blood. Should the retine concentration in blood be found to be low in cancer patients, then this might indicate that a low retine level predisposes one to cancer and metastases. By compensating for this lack by retine administration, cancer may, possibly, be prevented.

Various doubts may be raised about the therapeutic use of "retine." If retine does not harm the cell, but only prevents its division, how could one hope for a cure? In my experience the cancer cell is not a "wild-type" cell of increased vitality. It is a sick cell which has to rejuvenate itself by division to stay alive. If division is prevented for a long period of time, the cell has to wither away. Moreover, dividing cells may have an increased energy demand, and if cut off from their energy source by retine they would have to die.

Another criticism that could be raised concerning therapeutic application of retine is that since most of our cells have to divide now and then, cell division could not be completely arrested without danger. However, retine would, in all probability, have to be administered for a limited period only. Our experiments also indicate that cancer cells are much more sensitive to the action of ketone aldehydes than normal ones; the difference is considerable.

An objection that deserves careful consideration is that life depends on the undisturbed function of the hemopoietic apparatus, which could not be inhibited without fatal consequences. However, it seems likely to me that the regulatory mechanisms of the blood-forming apparatus are different from those of other tissues, and so may be insensitive to retine. Cell division in the liver or other tissues can be elicited by mechanical damage. The production of red blood corpuscles is regulated by the O_2 tension, whereas the production of white blood corpuscles may be elicited by antigens. These systems must be subject to different regulatory mechanisms, and so it is possible that retine will not act on these systems. It also follows that retine cannot be tested in leukemia.

Since Hammett (1929) called SH "The hormone of cell division," and since Rapkine's pioneering papers (1929, 1930, 1931, 1937) established a relation between SH-glutathione and cell division, a close relationship was suspected between SH and cancer. Retine may help to clarify these relations. The last review on this subject was published by Harrington (1967a,b), who also tried to link SH and cancer by a new theory.

It would be difficult for me to finish this chapter without paying tribute to Knock (1967) who, for many years, tried to direct attention to SH groups and their possible use in cancer therapy. My laboratory has derived more encouragement from the work of this brave pioneer.

The results I have obtained so far are only a hopeful beginning, not a goal achieved.

REFERENCES

Ashwood-Smith, M. J., Robinson, D. M., Barnes, J. H., and Bridges, B. A. (1967). *Nature (London)* 216, 137.

Bullough, W. S. (1962). *Biol. Rev. Cambridge Phil. Soc.* 37, 307.

Hammett, F. S. (1929). *Protoplasma* 7, 297.

Harrington, J. J. (1967a). *Advan. Cancer. Res.* 10, 247.

Harrington, J. S. (1967b). *Med. Proc. Meidese Bydraes* 13, 574.

Knock, F. E. (1967). "Anticancer Agents." Thomas, Springfield, Illinois.

Pullman, B., and Pullman, A. (1952). "Les theories électroniques de la chimie organique," pp. 171 and 172. Masson, Paris.

Pullman, B., and Pullman, A. (1963). "Quantum Biochemistry," p. 176. Wiley (Interscience), New York.

Rapkine, L. (1929). *C. R. Acad. Sci.* 188, 650.

Rapkine, L. (1930). *C. R. Acad. Sci.* 191, 871.

Rapkine, L. (1931). *Ann. Physiol. Physicochim. Biol.* 7, 383.

Rapkine, L. (1937). *J. Chim. Phys. Physicochim. Biol.* 34, 416.

Stich, H. F., and Florian, M. L. (1958). *Can. J. Biochem. Physiol.* 36, 855.